Praise for *Belo*

"A comprehensive history of the conservation movement—and a warning that we are not doing enough to prevent further animal mass extinction." —Barbara VanDenburgh, *USA Today*

"Excellent. . . . The book truly shines . . . when [Michelle] Nijhuis is brutally honest about how the conservation movement gained a reputation for being antihuman." —Sarah Zielinski, *Science News*

"With urgency, passion, and wit, Nijhuis . . . writes both to preserve history and predict what may lie ahead."
—Michael Berry, *Christian Science Monitor*

"Nijhuis is an engaging storyteller."
—Ernest Freeberg, *New York Times Book Review*

"Thoughtful and readable." —Andrew Robinson, *Nature*

"A far-ranging, powerfully written history of the conservation movement." —Alex Orlando, *Discover*

"Lavishly researched, *Beloved Beasts* is a compassionate look at what humans have done—and need to do next—to protect the natural world." —Amy Brady, *Literary Hub*

"Compelling and necessary reading for anyone interested in the field of conservation." —Rachel Love Nuwer, *Undark*

BELOVED BEASTS

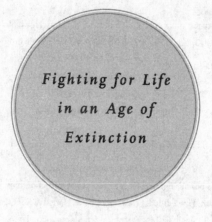

Fighting for Life in an Age of Extinction

MICHELLE NIJHUIS

W. W. NORTON & COMPANY
Independent Publishers Since 1923

For information about permission to reproduce selections from this book, write to
Permissions, W. W. Norton & Company, Inc., 500 Fifth Avenue, New York, NY 10110

For information about special discounts for bulk purchases, please contact
W. W. Norton Special Sales at specialsales@wwnorton.com or 800-233-4830

Manufacturing by Lakeside Book Company
Book design by Chris Welch
Production manager: Lauren Abbate

Library of Congress Cataloging-in-Publication Data

Names: Nijhuis, Michelle, author.
Title: Beloved beasts : fighting for life in an age of extinction / Michelle Nijhuis.
Description: First edition. | New York : W. W. Norton & Company, [2021] |
Includes bibliographical references and index.
Identifiers: LCCN 2020051063 | ISBN 9781324001683 (hardcover) | ISBN 9781324001690 (epub)
Subjects: LCSH: Endangered species—Conservation. | Wildlife conservation—History. |
Endangered species—Anecdotes. | Wildlife conservation—Anecdotes.
Classification: LCC QL81.7 .N55 2021 | DDC 591.68—dc23
LC record available at https://lccn.loc.gov/2020051063

ISBN 978-0-393-88243-8 pbk.

W. W. Norton & Company, Inc., 500 Fifth Avenue, New York, N.Y. 10110
www.wwnorton.com

W. W. Norton & Company Ltd., 15 Carlisle Street, London W1D 3BS

1 2 3 4 5 6 7 8 9 0

Maybe this is not a song about becoming extinct, though.
Maybe this is a song about becoming.

—E. M. LEWIS

Contents

BELOVED
BEASTS

AESOP'S SWALLOWS

When the swallow first noticed the mistletoe sprouting from the branches of the oaks, she gathered the other birds together, warning that humans would use a paste made from the mistletoe berries to trap them. Unless the birds could manage to tear the plants from the branches, she said, they should band together and beg the humans for mercy and refuge. The other birds laughed at the swallow, so she approached the humans alone. Charmed by her intelligence, the humans gave her sanctuary under their eaves.

Then, they proceeded to kill and eat the other birds.

This story is one of Aesop's fables—or, more precisely, it's said to be one of the fables told by an ancient wit called Aesop. Almost nothing is known about Aesop himself, and he may not have existed at all. If he did, he likely lived in the sixth century BCE. He was probably enslaved, possibly imprisoned, and perhaps taken into the confidence of his wealthy captors, his status elevated by his celebrated cleverness. Aristotle, born some two hundred years later, concluded that Aesop was Greek, but others thought he was Anatolian. Some said he was born mute; some said he solved the problems of kings; some said he was the companion of a famous courtesan, in spite of his resemblance to "a turnip with teeth." He told odd, unforgettable stories about tortoises and hares, lions and mice, boys and wolves. Legend has it that he talked his way out of trouble many times, until the day his tongue failed to save him and he was thrown off a Delphic cliff, much to the displeasure of the gods.

Generations of tellers have blended Aesop's fables with other stories, altering the morals to suit their own time and place. Present-day listeners will never know how his stories sounded when they were first told, or what they were intended to mean. What we do know is that they are rooted in the beginnings of what's often called Western civilization. And while Aesop's animal characters are usually proxies for humans, they are animal enough to show us that, even then, humans valued the members of other species in many ways: as individuals and types, calories and bait, companions and omens. Aesop's jokes lay in these contrasts, and his jokes were often on us.

What Aesop didn't know, couldn't know, was that more than two thousand years after his death, a few of the inheritors of his fables would recognize a more profound paradox in their own relationships with other species. The swallow, they would realize, was not the only kind of creature in need of sanctuary from themselves.

In the spring of 1994, when I had half an undergraduate degree in biology and precious few practical skills, I was hired as a field assistant on a wildlife research project in the desert of the southwestern United States. My job would have tickled Aesop; every morning before dawn, I hiked alone over the sandstone to a designated spot. There, I waited until the ground warmed and the tortoise I'd been assigned to watch dragged herself out of her burrow, blinking with what looked like weary exasperation. I spent the next few hours hovering at a distance, sneaking sips of water as I took notes on when and what the tortoise ate. When the day grew hot and the tortoise finally retreated into the shade, I stumbled back across the burning sand, fighting the feeling that, once again, I had somehow lost the race.

Desert tortoises were, and are, protected as a threatened species under the U.S. Endangered Species Act, and my weird, dreamy job was surrounded by bitter and sometimes violent arguments over their future. What struck me was how deep these arguments ran; while people fought over *how* to give sanctuary to tortoises—which construction projects or public-land grazing leases to curtail, which off-road vehicle tracks to close—they were also fighting about *why* and even *whether* they should be expected to protect tortoises at all. In the red rock canyons that once sheltered the Ancestral Puebloans, in meeting rooms and courtrooms and small-town coffee shops, people were arguing over nothing less than humanity's proper place on earth.

Left: Woodcut illustrating "The Swallow and the Other Birds," 1501. *Below:* Engraving illustrating "The Tortoise and the Hare," 1609.

After I returned to college and finished my degree, I spent several more years as an itinerant biologist. I searched California streams for endangered frogs and counted wildfire-charred plants in the Sonoran Desert, working with people who poked rattlesnakes for fun and called flowers by their Latin and Greek names. I learned about other species and my own, and later, as a journalist, I started to write about the relationships between them.

For more than twenty years, I've followed the arguments over the future of other species, and most of them sound familiar: we're still fighting over whether and why to provide sanctuary, and still waiting too long to figure out how. This stuckness has a lot to do with money and politics, of course, but it also has a lot to do with history—and with how we see (and don't see) that history today.

After all, it wasn't so long ago that many people in North America and Europe thought they understood their relationships with other animals, and with the groups of organisms called species. The Bible said that God granted humans dominion over the fish of the sea, the fowl of the air, and every living thing that moved upon the earth, and that more or less settled it—despite Aesop's rude reminders that the fish and fowl occasionally won.

In the middle of the nineteenth century, these complacent humans learned that they were both less exceptional and more powerful than they had believed. From Charles Darwin's theory of evolution, they learned that their ties to other animals were far closer than previously thought. They also came to realize that their rapidly industrializing and globalizing societies could drive species into extinction—and, in fact, had already done so, first on remote ocean islands and then much closer to home.

The most powerful, and powerfully destructive, civilizations on earth are still absorbing this double shock. No matter our creed or culture, most of us would agree that the animals we domesticate for companionship, labor, and sustenance shouldn't be abused (though our definitions of abuse may vary widely). Less clear are our responsibilities toward animals hunted for food and sport—and even murkier are those toward animals who are annoying, dangerous, or allegedly useless. The understanding that humans are capable of exterminating entire groups of organisms raised a new question: Why should any of us make sacrifices, even in the short term, to ensure the persistence of other species on the planet?

Until very recently, Western philosophers and religious scholars paid little attention to such questions, and many scientists shied away from them. Laws that try to answer them often raise more dilemmas than they settle.

And Aesop—well, Aesop would probably laugh and wish us luck.

This book is about the humans who have devoted their lives to these questions—the scientists, birdwatchers, hunters, self-taught philosophers, and others who have countered the power to destroy species with the whys and hows of providing sanctuary. Each person profiled here stood, or stands, at a turning point in the story of modern species conservation—a story which, for better and sometimes worse, still guides the international movement to protect life on earth.

Most early conservationists were privileged North Americans and Europeans, and no wonder; location and education enabled them to recognize the effects of humans on other species, and money and status freed them to take controversial positions. While the societies around them were busy forgetting their dependence on the rest of life, they were forming a new kind of attachment to it, caring not only about individual animals but about the survival of species.

Though they often used pragmatic arguments to convert others to their cause, their personal motivations ran deeper, for many had started keeping company with members of other species to escape their own troubles. Some were painfully shy, or burdened with mental or physical illness. Some were separated from spouses at a time when divorce was a scandal, or drawn to their own gender when homosexuality was taboo. Most of them knew something about suffering, and they found consolation in the sights and sounds of other forms of life.

Their passion for species often began with colorful birds or large, showy mammals, but it grew to include the tiny, the unknown, the stationary, and even the despised—and, importantly, the relationships among them all. More often than not, these conservationists failed to save the kinds of life they loved; they were defeated by indifference, their own blind spots, or the human instinct for increase. But more often than many of us realize, they succeeded, and their intellectual descendants continue to succeed today.

The story of modern species conservation is full of people who did the wrong things for the right reasons, and the right things for the wrong reasons. It begins in wealthy countries and in colonized territory; its early chapters are shadowed by racism, and some conservationists still hold blinkered views of their fellow humans, causing them to mislay blame for the damage they seek to contain. Many conservationists are unfamiliar with their movement's history—as geographer William Adams observes, they tend to ricochet between evolutionary time and the crisis of the moment—with the result that each generation has revived old arguments and repeated mistakes.

Yet what began as a series of running battles to protect charismatic animal species has developed, over the past century and a half and through countless twists and turns, into a global effort to defend life on a larger scale. Now, as the destruction of species continues and the effects of climate change escalate, conservationists worldwide are fighting for the future of all species, including our own.

To consumers of modern media, the story of species conservation doesn't look much like a story. It looks like a jumble of tragedies and emergencies: the last Yangtze river dolphin, the last two northern white rhinos (both female), the soon-to-be-last vaquita—obituaries and near-obituaries relieved only by sporadic heroics and temporary successes. It takes place in a bleak present and a much bleaker future. It tempts us with the fantastical (resurrected herds of woolly mammoths) and the exceptional (extinction averted one costly, awkward artificial insemination at a time). Perhaps most dangerously, it tempts us with despair.

It's easy to forget that the world we live in is far richer thanks to those who found convincing reasons, and the required means, to provide sanctuary to other species. Without their work, there would likely be no bison, no tigers, and no elephants; there would be few if any whales, wolves, or egrets. And there would likely be no international tradition of conservation. Over decades of alliances and arguments, conservation has expanded beyond its privileged beginnings, establishing a movement that is shaped by many people, many places, and many species.

We can learn from this tradition: from its successes and failures, its oversights and insights. For while new technology, from drones to cryopreservation to gene editing, may well assist future conservationists, there is still no shortcut through the difficult work of protecting living organisms and the places they need to survive. And there is still no substitute for the emotional connection with other animals and other species—the love of life that inspired the first modern conservationists and continues to sustain so many today.

Humans and their fellow animals are indeed facing novel and pressing problems. The climate is changing as never before in human history. Frogs, bats, and salamanders are suffering their own pandemics, dying from diseases spread by human activities. An estimated one million species are now threatened with extinction, including as many as a quarter of all animal and plant species. Organized crime and entrenched corporate interests threaten conservation efforts and, in many places, conservationists themselves; between 2002 and 2019, at least eighteen hundred people were murdered in the course of defending land, water, plants, and animals from poaching and other human insults.

Fantasy and despair are tempting, but history can help us resist them. The past accomplishments of conservation were not inevitable, and neither are its predicted failures. We can move forward by understanding

the story of struggle and survival we already have—and seeing the possibilities in what remains to be written.

In its broadest sense, conservation is the prevention of waste or loss.
Measures to prevent the waste or loss of certain kinds of fish, birds,
and mammals—for spiritual, practical, and self-interested reasons—
are likely as old as the images of steppe bison painted on cave walls by
Paleolithic hands, and they have been (and still are) employed by societies large and small, poor and powerful.

The Indian emperor Ashoka, who converted to Buddhism in the third
century BCE, forbade the killing of parrots, tortoises, porcupines, bats,
ants, and "all four-footed creatures that are neither useful nor edible."
Marco Polo reported that Kublai Khan, who ruled the Mongol Empire
in the thirteenth century, prohibited the hunting of hares, deer, and
large birds between March and October "that they may increase and
multiply." Under the draconian forest-law system imposed by medieval
English kings, hunting was reserved for royal pleasure and royal profit.

The people in this book saw the need for a more specific kind of
conservation—a sustained international campaign to protect other
species from human-caused global extinction. While the conservation
movement has variously ignored, disrupted, and idealized older forms
of conservation, recent generations of conservationists are recognizing
that their work is part of what their predecessor Aldo Leopold called
"the oldest task in human history," and are striving to synthesize the
best of old and new.

Early conservationists were divided by arguments between utilitarians, who were primarily interested in sustaining herds and forests for
human use, and preservationists, who mostly wanted to protect species
and landscapes from human interference. Since then, the conservation movement—whose founders included many wealthy sportsmen—
has sometimes come into conflict with the environmental movement,
which was sparked by public concern about global problems such as air
and water pollution. The animal welfare movement, which began as an

effort to improve the lives of domesticated animals, has been ally and antagonist to both conservation and environmentalism. Yet all of these movements are closely related, and their stories overlap.

Today, conservation is defined in numerous ways—it can include, for instance, the protection of scenic vistas and open space—but the persistence of species, and of animal species in particular, remains a vital concern of the movement, and the history and future of that concern is the focus of this book.

Conservation has long been a collaboration between passionate experts and passionate amateurs, and it blows through disciplinary boundaries, drawing from science, politics, law, philosophy, and other fields. Its internal debates are lively and often fascinating, but they are almost always complicated by the use of terms that have come to mean very different things to different people. Conservation is complicated enough, so I've minimized my use of words such as "nature," "wild," and "wilderness." When I do use them, I've tried to make clear which nature, or which wilderness, I'm talking about. I use "wildlife," another term with multiple definitions (and various orthographies), simply to distinguish nondomesticated animals from their domesticated counterparts, including human beings. In general, the terms I use to identify humans—Victorian, Blackfeet, white, Buddhist—are the terms used by the people themselves.

There's one more word I use sparingly here: hope. Hope is the subject of much discussion in conservation circles, both the need for it and the lack of it. Yet few if any of the most influential early conservationists were motivated by what might be called hope. They were motivated by many other things—delight, outrage, data—but they had little confidence that the work they were moved to do would succeed in rescuing the species they loved. They did it anyway. As Leopold, in one of his grimmer moods, wrote to a friend, "That the situation is hopeless should not prevent us from doing our best."

Chapter One

THE BOTANIST WHO NAMED
THE ANIMALS

On the southeastern outskirts of Washington, DC, across the Anacostia River from the marble monuments of the national capital, lies the Smithsonian Institution's Museum Support Center, a vast concrete zigzag which holds the institution's undisplayed collections. Inside are some fifty-five million objects, among them relics of wonder and discovery, slaughter and subjugation: meteorites recovered from Antarctica; the skulls of four elephants shot by Theodore Roosevelt during his African safari in 1909; a fifty-two-foot-long dugout canoe, its hull formed from a single red-cedar tree, built by the Nuu-chah-nulth people of Vancouver Island.

Here, too, are thousands of type specimens—the individual plants and animals that serve as examples of their species. Every time scientists describe and name an unfamiliar species, they collect or capture at least one representative, then skin, dry, or otherwise preserve it and store it in a museum for posterity. This taxonomic tradition dates back more than a century, and so do some of the specimens in the Museum Support Center.

Most of the Smithsonian's reptile and amphibian specimens, type and otherwise, are stored in the center's Pod Five, a high-ceilinged, blindingly white room full of rolling steel shelving units. On the immaculate shelves stand rows and rows of jars, each filled with alcohol and one or more fleshy, bloodless bodies.

Steve Gotte cares for this collection, a job that initially struck me as peaceful, maybe even meditative. But Gotte, his black mustache obscuring any hint of a smile, quickly informed me otherwise. The jars left open by careless researchers, threatening to turn specimens into mush! The substandard rubber seals that melt like candle wax! The less-than-meticulous field notes, filled with obscure locations! "It's all cool," commented a lab technician, as he scanned an almost unreadable page from a Victorian-era ledger, "but sometimes you just want to smack these guys."

In recent decades, genetic analysis has made it possible for scientists to recognize so-called cryptic species, species whose types look identical to others but are genetically distinct. Each new species designation requires jars to be refilled, relabeled, relocated. Gotte gestured toward his desk at a new taxonomic monograph the size of a small telephone book. "There used to be one species of slimy salamander," he said. Now there are sixteen.

The shelves of Pod Five record other changes, too. Gotte showed me specimens of *Incilius periglenes*, the golden toad of Costa Rica, a species thought to have been extinguished in the 1980s by *Batrachochytrium dendrobatidis* or *Bd*, a fungus that originated in Asia and spread worldwide via international trade. So far, *Bd* is estimated to have contributed to the decline or extinction of five hundred amphibian species—more than 6 percent of the known global total—making it by some measures the most destructive pathogen on earth.

I picked up a jar of *Atelopus ignescens*, commonly known as the Jambato toad or Quito stubfoot toad. Coal-backed and orange-bellied in life, the colorless, inch-long bodies stirred gently in their alcohol bath. These toads were once so abundant in the Ecuadorian highlands that people had to shoo them out of their houses, and these particular specimens, collected during a roadside stop on an August day in 1962, were probably an unremarkable find.

A. ignescens remained unremarkable for the next quarter century, but in the late 1980s the species called attention to itself by going missing, even from places apparently undisturbed by humans. Despite regular surveys by scientists, it was not documented again until 2016, when a young boy discovered a tiny remnant population. (In biological terms, a population is generally defined as a group of individuals from the same species living in the same place at the same time.) Today, there are almost certainly more *A. ignescens* living in captivity than at large in the Andes, and related species are in similar straits.

I struggled not to look sentimental in front of Gotte, and wondered why I was moved. I've always been partial to frogs and toads, for many reasons and none at all. I like them because they remind me of childhood hours spent playing in muddy ditches; because they can live in two different worlds; because they're ugly and beautiful, fragile and resourceful. ("Frogs are strange creatures," Henry David Thoreau mused in 1858. "One would describe them as peculiarly wary and timid, another as equally bold and imperturbable.") I like them because a lot of other people don't, and I also just like them, in the way that some people like the color blue. I often hear scientists express the same sort of instinctive inclinations—and disinclinations, for that matter. I once met a marine biologist who was fascinated by the most grotesque kinds of sea life, but found songbirds "too twitchy."

But why do I care about the fate of this particular species? The quiet disappearance and equally quiet reappearance of *A. ignescens* hasn't affected me directly, at least not in any way I know of. I live more than four thousand miles away from the Andes, in the northwestern United States, where I can see plenty of frogs and toads any time I want. Part of what gets me about *A. ignescens*, I think, is the speed and recency of their exit. When I was in elementary school, these toads were common. By the time I got to high school, they were believed to be gone for good, and they remain vanishingly rare. Like most everyone else who has peered at these pickled toads, I also understand that they represent a distinctive type of life, likely the result of millions of years of evolution.

I have a sense that their kind can't be replaced, and will be missed in immeasurable ways. And what gives me that sense is their unique scientific name: *Atelopus ignescens*.

Unlike common names, which tend to change over time and are often shared by more than one species, that two-part scientific name identifies this type of toad as not only distinctive in its neighborhood, but distinctive in the world. At the same time, it locates *A. ignescens* on the family tree of life, establishing its relationship with better-known species and turning what might otherwise be distant abstractions into real creatures. The names we choose for plants and animals have long been fundamental to our understanding of the world around us—and in many places, elaborate local taxonomies are still in service. While scientific names have sometimes been used to erase these older systems of designation, they can also stitch them together, enlarging our collective view of the world and its living variety.

The biologist E. O. Wilson has said that the purpose of taxonomy is to "find out what the animals themselves know"—to group creatures as they group themselves, in other words, and treat them accordingly. We will never know what the animals know; to extend Wilson's conceit, our names for them will never be their names, and our presumptions about their experiences will never be their experiences. Taxonomy, however, can give us a distant sense of their worlds, and—ever more importantly—a means of both recognizing their different kinds and knowing when those kinds are in decline. For that, we can thank the generations of taxonomists who have formally named about two million of the estimated nine million species on earth. And all of them can thank one particular Swedish botanist, whose three-hundred-year-old corpse is the unofficial type specimen for *Homo sapiens*.

If you had bumped into young Carl Linnaeus in the muddy streets of Uppsala, Sweden, in the spring of 1729, he probably would have told you he was destined for great things. You probably wouldn't have believed him.

His parents, crushed by his decision not to follow his father into the Lutheran Church, had sent him off to Uppsala University the previous August with a single pocketful of silver, which he had long since spent on food and fuel. The winter had been unusually severe, and Linnaeus, who loved the outdoors but hated the cold, had shivered through the season in rags and paper-soled shoes. The university, though the oldest in Scandinavia, was as shabby as Linnaeus himself; almost obliterated by a fire a generation earlier, it still had no chemistry laboratory, only a makeshift anatomy laboratory, and a botanical garden that Linnaeus found hardly worth visiting. Professors offered costly private lessons in place of public lectures, and students started drinking soon after midday, sometimes finishing their benders by assaulting unpopular instructors.

Linnaeus, though, was already gripped by a grand ambition. Since early childhood, he had been fascinated by plants, and his father, an avid gardener, had introduced him to the flowers around their home in southern Sweden. Linnaeus, whose schoolmates had nicknamed him "the little botanist" for his habit of cutting class to roam the woods, spoke of plants as if they were people, and treated them as friends. Now, he was determined to name and order them all. "From the depths of his heart he detested flux and formlessness," writes Swedish historian Sten Lindroth. "He had come to the world to tidy up and set in order."

Like other boys of his time and place, Linnaeus had been schooled in the works of the ancient Greeks, absorbing the Aristotelian view of a world composed of only a few hundred types of plants and animals. He gained his "first view of the conveniences of arrangement and the beauty of system," as a Victorian biographer put it, from the work of seventeenth-century French naturalist Joseph Pitton de Tournefort, whose three-volume encyclopedia of plants classified about ten thousand species. He knew that the English parson John Ray—the first naturalist to define a species in biological terms—had recently described eighteen thousand plant species. He also would have heard that European traders and colonists were busily bringing unfamiliar species back from what was known as the New World, and that the haphazard naming of these plants and animals was leading to colossal confusion—and rampant fraud—in gardening and natural history circles. Even so, a world containing millions of distinct plant and animal species would have been almost impossible for Linnaeus to imagine. Luckily for him, and for science, he was naïve enough to believe that his dream of ordering life was achievable—and confident enough to believe that it was his to achieve.

Linnaeus soon found his footing at Uppsala. He befriended a fellow medical student, Peter Artedi, who shared his zeal for classification, and they divided the living world between themselves. Artedi, who inclined toward zoology, claimed the amphibians, reptiles, and fish; Linnaeus claimed the birds, insects, and most of the plants; mammals and minerals were common territory. Artedi, two years older than Linnaeus, was in many respects his antithesis: tall, thin, and dark-haired where Linnaeus was short, broad, and fair; earnest and reflective where Linnaeus was bold and mercurial; modest where Linnaeus was anything but. Together, they searched the woods and meadows around Uppsala for species to catalog, one or the other sometimes hiding a discovery for days before breaking down and bragging about it. As Linnaeus recalled, they vowed that if one of them died, "the other would regard it as a sacred duty to give to the world what observations might be left behind by him who was gone."

Their fervent collaboration was short-lived. Linnaeus's reputation as a botanical prodigy came to the attention of a university dean, leading to a succession of important associations for the young scholar. By 1735 he was living in the Netherlands, in the university city of Leiden, and talking his way into an audience with the renowned Dutch physician and botanist Herman Boerhaave. One morning, in a tavern, Linnaeus spotted a familiar face; Peter Artedi had come to the Netherlands, too. In his memoirs, Linnaeus made no mention of why he and Artedi had lost touch, but recalled that the chance reunion moved both to tears.

By all accounts, Artedi was at least as gifted and passionate about his work as Linnaeus, but he seems to have lacked his friend's willingness to ingratiate himself with the powerful, and his fortunes suffered for it. After graduating from Uppsala, he had studied fish classification in London, working with the best in the field. When his money ran low, he had come to Leiden for further study, but the cost of the trip left him in desperate straits. Linnaeus found Artedi some extra shirts and, before long, a job, arranging for him to work with a well-known apothecary in nearby Amsterdam.

Linnaeus went to work in Amsterdam as well, but Artedi, always reserved, withdrew further into drink and work. One night, Linnaeus visited his rooms to show him a new manuscript, and Artedi, in turn, insisted on reading aloud his entire work-in-progress on ichthyology and chewing over its finer points. "He kept me long, too long, unendurably long (which was unlike our usual practice)," Linnaeus wrote, "but had I known that it was to be our last talk together I would have wished it even longer."

Carl Linnaeus and Peter Artedi, as imagined by an illustrator in 1878.

On September 27, 1735, Artedi attended a dinner party given by his employer, and stayed until well after midnight. On his way home, he stumbled into an unfenced canal and drowned. He was thirty years old. When Linnaeus got the news, two days later, he rushed to the city hospital to find, as he remembered, "the stiff and lifeless body, the livid, pale and foam-flecked lips." Only ten weeks had passed since the two reunited in Leiden.

Linnaeus had not forgotten his pledge to share his friend's work with the world, but Artedi was so behind on rent that his landlord confiscated his papers as collateral, and Linnaeus had to borrow money from his latest sponsor to buy them. He discovered that Artedi, despite his troubles, had succeeded in ordering all known fishes into nested categories, and describing each species in elegant detail. Artedi had not achieved his and Linnaeus's shared dream, but he had made more progress toward it than his better-appointed friend. With what must have been a mixture of admiration and envy, Linnaeus wrote of Artedi's work to a colleague, "You will see more perfection than can be expected in botany for a hundred years to come."

Artedi was buried in an unmarked pauper's grave in Amsterdam, but he was survived by his book *Ichthyologia*, which Linnaeus published in 1738. And his work lives on, invisibly embedded into every scientific name in use today.

After his sojourn in the Netherlands, Linnaeus returned to Uppsala, where he married, fathered seven children, and settled, with apparent satisfaction, into a professorship at the university. He rarely left Sweden again. His "apostles," a corps of former students with more tolerance for the discomforts of adventure than their professor, set out on colonial expeditions, returning with unfamiliar plants that Linnaeus added to his jealously guarded collection. ("Let no naturalist steal a single plant," he instructed his wife before his death.)

Though Linnaeus gained early fame—wealthy admirers helped him publish a preliminary version of his masterwork, *Systema Naturae*, in

1735, while he was still living in Leiden—his ideas were far from universally popular. His proposal to organize flowering plants according to the characteristics of their sexual organs scandalized botanists throughout Europe, many of whom considered his mischievous descriptions of flowers as having "two husbands in the same marriage" or "twenty males or more in the same bed with the female" altogether too memorable. "To tell you that nothing could equal the gross prurience of Linnaeus's mind is perfectly needless," the Rev. Samuel Goodenough wrote to a Linnaean scholar in 1808, before conceding that "It is possible that many virtuous students might not be able to make out the similitude of *Clitoria*." Other critics accepted Linnaeus's metaphors but not his methods; they simply didn't want to change their own systems, unwieldy as they might be.

The Linnaean naming system proved to be an irresistible innovation. Naturalists were already using two-part names to sort organisms into categories and subcategories, but Linnaeus formalized this practice and placed it within the nested hierarchy Artedi had established for fish. The result was an efficient, adaptable mechanism for ordering life on earth.

In *Species Plantarum*, published in 1753, Linnaeus reduced the cumbersome descriptive names used for many plant species to tidy Latin and Greek binomials: *Convolvulus foliis palmatis cordatis sericeis: lobis repandis, pedunculis bifloris* became *Convolvulus althaeoides*, which is still the scientific name of the pink-flowered morning glory commonly known as mallow bindweed. In the tenth edition of *Systema Naturae*, published in 1758, he rechristened much of the rest of the known living world, including humans, and assigned each species a place in an expanded classification system that began with kingdoms and, in its modern form, branches into phyla, classes, orders, families, genera, and finally, species. (The classic biology student mnemonic for this hierarchy is King Philip Came Over From Genoa Spain, though Philip sometimes Comes Over For Good Spaghetti or Great Sex). Following the practice he established in *Species Plantarum*, Linnaeus formed the name of each species in *Systema Naturae* from its genus and its specific epithet.

Left: Linnaeus after his 1732 expedition to northern Sweden, wearing Sami dress and holding his favorite plant, a twinflower (named *Linnaea borealis* in his honor). *Right:* Linnaeus in later life, again holding a twinflower.

By the time Linnaeus died, in 1778, he had coined unique, internationally understood names for more than four thousand species of animals and nearly eight thousand species of plants. In Sweden, he was honored on coinage and in statuary, and throughout Europe he was known as the "Prince of Flowers" and the "King of Botany." He achieved, as historian Donald Worster writes, "elementary order for an era of anarchy in natural history."

Many of the species Linnaeus described have since been renamed, and some have been determined not to exist at all—such as the phoenix and the hydra, which he placed in a group called *Paradoxa,* and the kingdom of minerals that he mystifyingly included in successive editions of his taxonomy (and which survives today as a category in the "animal, vegetable, mineral" guessing game). Most notorious are the four "subspecies" of humans that Linnaeus described by skin color, a taxonomic fallacy that haunts us still.

But more than five thousand plant and animal species are known to science by the same names Linnaeus gave them. Some are named for people he wanted to flatter, and a few, such as the smelly little weed *Sigesbeckia*—named for a particularly harsh and persistent critic—are his enduring revenge.

Despite his self-regard, Linnaeus realized that no single mortal could achieve his boyhood goal of naming and classifying all of life on earth. He published detailed instructions for his fellow taxonomists, and in 1895, many of his principles were adopted by the International Commission on Zoological Nomenclature, the organization that monitors the christening of all newly identified animal species. (The International Association for Plant Taxonomy does the same for algae, fungi, and plants, and there are equivalent associations for bacteria and viruses.) "Ordinary languages grow spontaneously in innumerable directions; but biological nomenclature has to be an exact tool that will convey a precise meaning for persons in all generations," declares the ICZN code, which was formalized in 1961.

While the ICZN established and maintained the rules of biological nomenclature, it didn't start keeping track of the world's existing scientific names until 2012. For more than a century, that job fell to people like Charles Davies Sherborn, a London bookseller who spent forty-three years extracting almost half a million species names from tens of thousands of published sources. Sherborn's *Index Animalium*, completed in 1933, is thought to include all of the animal species, living and extinct, named after the 1758 edition of *Systema Naturae* and before 1850. Digital technology has since eased some aspects of indexing—it took Sherborn three years just to alphabetize his list—but taxonomists are still plagued by duplicate species names, unorthodox spelling, and other very human errors.

Contemporary taxonomists can't hope for the glory Linnaeus enjoyed. They aren't immortalized as statues or on coins, and the work of naming and describing species—not to mention indexing them—is chronically underfunded. E. O. Wilson, among others, has decried this decline, calling for a "renaissance" dedicated to "the description and mapping of the world biota"—including the millions of species still entirely unknown to science.

Wilson's interest is both intellectual and practical. As Linnaeus observed, "If you do not know the names of things, the knowledge of them is lost too." Species without formal names aren't surveyed by researchers, and their needs aren't systematically assessed. They can't be protected by the U.S. Endangered Species Act, or added to the international Red List of Threatened Species. They might be loved and appreciated by humans, but only in the immediate neighborhood—not from any distance. "For in order to care deeply about something important," Wilson writes, "it is first necessary to know about it."

Scientists do continue to name unfamiliar creatures with precision and creativity. *Atelopus ignescens*, the species name of the toads preserved in Pod Five, was inspired by their fire-orange bellies. A related toad, identified in 2004, when it was already critically endangered, was named *Atelopus epikeisthos* after a Greek word meaning "threatened through adverse circumstances." (The genus name *Atelopus*, coined by two French zoologists in the 1840s, is based on the Greek words for "imperfect" and "foot," likely because the toads in the genus have distinctively short first toes.)

Every year, about eighteen thousand species join the Linnaean ranks, and in 2020 they included a newly recognized species of *Smaug*, a genus of southern African lizards named after the "most specially greedy, strong and wicked" dragon in *The Hobbit*; three species of soft-nosed chameleons from Madagascar, distinguished from one another in part by small differences in their branched penises; and the praying mantis *Vates phoenix*, described with the help of specimens borrowed from Brazil's National Museum before the building was devastated by fire in late 2018. Many newly identified species are reported by dedicated amateurs—retirees, pediatricians, bus drivers, and others who spend their spare time exploring little-studied genera of invertebrates and fungi.

Even in well-trafficked places, it's still possible to find unnoticed forms of life. During a reporting trip to Cambodia in 2012, I met a young British ornithologist who had spotted an unusual orange-capped bird on the floodplains around the capital city of Phnom Penh. After weeks of looking and listening, he and his Cambodian colleagues confirmed that the bird was a member of a scientifically unrecognized urban species. Following Linnaean rules, they named it the Cambodian tailorbird, *Orthotomus chaktomuk*.

Professional etiquette prevents scientists from naming species after themselves, but they have enthusiastically continued the Linnaean tradition of naming species after friends, enemies, and heroes. There is a species of horsefly named after Beyoncé (*Scaptia beyonceae*), a spider after David Bowie (*Heteropoda davidbowie*), a termite after the Colombian artist and political satirist Fernando Botero (*Rustitermes boteroi*), a tiny neon-purple fish after the fictional African nation ruled by the Black Panther (*Cirrhilabrus wakanda*), and a millimeter-long tropical beetle after the young Swedish climate activist Greta Thunberg (*Nelloptodes gretae*).

Though the ICZN code forbids the coining of names that would be "likely to give offence" to their eponyms or the public, there are species of slime-mold beetles named after former U.S. president George W. Bush and his vice president, Dick Cheney. In early 2017, a moth with feathery yellow head scales was named *Neopalpa donaldtrumpi*.

Since there is no procedure for changing scientific names that *do* give offense, one eyeless, cave-dwelling Slovenian beetle species retains the name *Anophthalmus hitleri*—a burden it acquired in 1937, courtesy of an entomologist who may or may not have wanted to flatter the Führer. (Hitler reportedly interpreted the recognition as a compliment, and sent a letter of thanks.) Today, specimens of *A. hitleri* sell to neo-Nazi collectors for four figures a pop, and poaching pressure is so intense that the species is nearly extinct.

Most of the type specimens for these and other species are stored in places like Pod Five, but at least one has a very different kind of resting place. In 1959, in an essay about the legacy of Linnaeus, British biologist William Stearn noted that *Homo sapiens* had no designated type. Stearn suggested that Linnaeus, who had named the species and was so determined to preserve himself that he wrote his own autobiography five times, might appreciate the honor. Linnaeus, unlike so many of his fellow types, was not suspended in alcohol but allowed to remain where he was buried—beneath a gravestone set into the floor of Uppsala Cathedral, memorialized as an example of his species.

In their battle against flux and formlessness, Linnaeus and other early taxonomists were more concerned with easing communication among naturalists than with reflecting the real-world relationships among different kinds of organisms. To them, the metaphorical tree of life was more like a set of pantry shelves, whose categories were defined not by kinship or even overall resemblance but by a few isolated, conveniently observable characteristics.

Linnaeus lived more than a century before Charles Darwin published his theory of evolution, and he believed that species were enduring entities created by God. "*Natura non facit saltus,*" he wrote in *Philosophia Botanica*; Nature doesn't jump. (Linnaeus apparently became less sure about this later in life, for he crossed out the line in his own copy of the book.) His thesis *Oeconomia Naturae,* written in 1749, is a forerunner of ecological thought, describing living organisms as "so connected, so chained together, that they all aim at the same end." But there was no scarcity in this divinely conceived economy, according to Linnaeus, and thus no need for species to change.

Darwin realized that resources were indeed limited, and that each type of life was made up of a variety of individuals who competed with one another. His understanding that the more successful individuals in a population would be more likely to pass on their advantages, causing both populations and species to merge, diverge, and transform over time, turned the tree of life into a *family* tree. German zoologist and artist Ernst Haeckel, whose famous 1874 illustration represented evolutionary history as a gnarly oak, planted the family tree of life in the public imagination—and, by placing *Homo sapiens* at its crown, saddled Darwin's theory with Haeckel's own sense of human destiny.

In the decades after Darwin, the Linnaean search for order became a search for origins. As scientists began to understand how individual

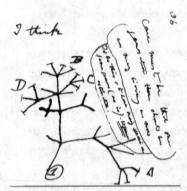

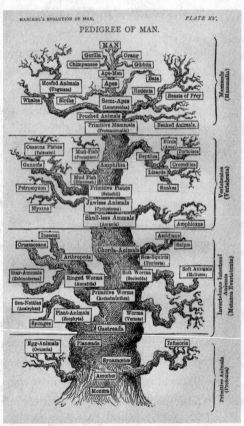

Above: Charles Darwin's 1837 sketch of an evolutionary "tree" shows evolution proceeding in many directions. His note reads: "I think . . . case must be that one generation should have as many living as now. To do this and to have as many species in same genus (as is) requires extinction." *Right:* In Ernst Haeckel's tree of life, published in 1879, evolution progresses toward "man."

traits passed from one generation to the next, taxonomists used these traits to classify species by their presumed evolutionary lineages. The nested hierarchies of the Linnaean system were well-suited to this new approach, and they weathered the transition more or less intact. During the second half of the twentieth century, however, the ability to trace inherited sequences of genetic code raised new questions—not only about the arrangement of species on the tree of life, but about the very nature of a species.

In the late 1970s, genetic analyses of what appeared to be an unusual type of bacteria revealed the archaea, a previously unrecognized category of life so broad, and so distinctive, that its classification required the creation of a new taxonomic level, the domain. Today, the three generally accepted domains of life are the Eukarya, which includes the plant and animal kingdoms; the Bacteria; and the Archaea. (The King Philip of the student mnemonic was updated to Dear King Philip.) Researchers have come to realize that archaea—as well as other, more familiar forms of life—can pass genetic information not only from one generation to the next but also within generations, through a process called horizontal gene transfer.

These and other discoveries challenged the widely accepted notion of a species as a group of organisms capable of interbreeding, and led scientists to propose dozens of finer-grained definitions. (Historian and philosopher of science John Wilkins offers this type specimen of nerdy taxonomy humor: in a room of n biologists, one can find $n + 1$ definitions of a species.) While the ICZN and like organizations govern how species are named, there is no ultimate authority on what constitutes a species; as the caretakers of Pod Five can attest, new genetic information almost invariably kicks off a taxonomic quarrel.

Some scientists, nicknamed splitters, argue for new species to be recognized based on subtle genetic or physical differences, while others—the lumpers—argue for keeping the number of species as small as possible. While these debates may sound esoteric, they can have major consequences for conservation. In 2016, after years of analysis and argument, researchers concluded that there are four species of giraffe, not just one, and that at least two of the four species are in danger of extinction. Meanwhile, taxonomists have proposed to reduce the number of recognized tiger subspecies from nine to two, distinguishing only between tigers on the Asian mainland and those on the islands of Indonesia—raising the possibility that tigers from the relatively abundant population in Siberia could be introduced into the critically endangered population in China. When ornithologists conducted a review of bird taxonomy in the mid-2010s, they lumped some species together and split many more, ultimately recognizing more than a thousand new species worldwide. Some are more endangered than their antecedents, but others are less so.

Taxonomies have always been human constructs, influenced by history and politics, geography and grudges. But the living organisms that underlie them are biological realities, and species, disputed boundaries and all, reflect real associations and distinctions. No one needs a genetic test to know that a lion is different from a jaguar, or that a zebra is different from a donkey. Biologists may one day decide to lump *Atelopus ignescens* with another species, or split it into three, but there's no question that the pickled individuals in Pod Five are both related to and gloriously, poignantly distinct from every other kind of toad I've had the chance to see up close. Thanks in large part to the clarity of the Linnaean naming system, our fuzzy-bordered categories for biological realities have also become political, legal, and emotional realities. And in all of those senses, species are the fundamental units of modern conservation.

What we *Homo sapiens* are still struggling to accept is that we can obliterate so many of them, and in so short a time.

Chapter Two

THE TAXIDERMIST
AND THE BISON

n late February of 1888, spring was starting to creep toward Washington, DC, but on the southern edge of the National Mall, William Temple Hornaday was conjuring the slicing chill of the Montana plains. Inside a newly built red-brick building, behind a set of privacy screens, Hornaday knelt on the floor of an enormous glass and mahogany display case, gluing down sprigs of buffalo grass and sage and scattering handfuls of genuine Montana Territory dirt.

Surrounding him on the simulated scrap of prairie were six shaggy figures, the reconstructed remains of the bison he and his companions had pursued and killed during a pair of grueling expeditions two years earlier. The largest animal, a bull Hornaday shot himself at the end of the second trip, had measured more than nine feet from end to end, weighed some sixteen hundred pounds, and had six old bullets buried in his flesh. Hornaday was a relatively small man, about five foot eight, and when he was at work inside the case, overshadowed by the remnants of his quarries, he must have looked positively puny.

Hornaday was the chief taxidermist at the United States National Museum, the institution that would soon be known as the Smithsonian National Museum of Natural History—and whose holdings now overflow into the Museum Support Center. Taxidermy, in Hornaday's

William Temple Hornaday (center) and two colleagues in the taxidermists'
laboratory of the United States National Museum, circa 1880.

time, was a respected and even lucrative occupation for those inclined
to adventure; bird watching, shell collecting, and other types of "nature
study" were fashionable in North America and Europe, and, since zoos
were rare, museums were eager for artfully preserved specimens. Hornaday had spent his twenties traveling the world as a trophy hunter, battling crocodiles and life-threatening illnesses in order to bring exotic
remains home for museum display. At thirty-three, he was at the top
of his profession, and consumed by an intensely ironic new mission.

During their expeditions to Montana Territory, Hornaday and his
companions had killed more than twenty of the continent's last free-
roaming bison—all of them American plains bison, a subspecies
known to science by the insistent Linnaean trinomial *Bison bison bison*.
Under the starry prairie sky, Hornaday had dismembered the carcasses

and drained their blood, carefully cleaning and preserving the hides and bones for their journey east. Cloistered in the museum, he was engrossed in what one reporter called "about the biggest thing ever attempted by a taxidermist." And he was determined that the heavy hides of his six sacrificial beasts, stretched over clay-covered wooden models and posed on an imitation prairie, would protect the bison who had evaded his bullets.

Toward the northwestern corner of Montana, above the confluence of the Sun and Missouri rivers, a mile-long sandstone cliff rises out of the prairie like a frozen wave. Today, the ridge is the centerpiece of First Peoples Buffalo Jump State Park. (American bison were nicknamed "buffalo" by early European settlers, who may have noticed the animals' passing resemblance to African Cape buffalo and Asian water buffalo, and the informal name stuck.) During Hornaday's time and for centuries before that, this place was known to many as a *pishkun*, a Blackfoot word that loosely translates to "deep kettle of blood."

On the late-spring morning that I visited the park, a stiff wind was pouring off the eastern slopes of the Rocky Mountains, and the sun glared out of a cloudless sky. Montana was just emerging from a winter of record-setting snows, but spring is arriving earlier and faster as the climate changes, and the sudden rush of snowmelt had briefly turned the prairie green.

Near the foot of the shallow slope that rises to the cliff edge, I met Lyle Heavy Runner, a Blackfeet tribal member who grew up on the tribe's reservation in northwestern Montana. After a peripatetic career with FedEx, he lives in the city of Great Falls, a few miles from the state park. We drove up the winding road that leads to the top of the ridge, then walked to the precipitous edge. The Sun and Missouri rivers glittered in the distance. Below our toes was nothing but wind. Heavy Runner shuddered. "I hate heights," he said.

American bison are not easy to kill. Both subspecies of *Bison bison*— *Bison bison bison* and its larger northern cousin, *Bison bison athabas-*

cae, the wood bison—are huge animals, and male wood bison can be nearly as big and hefty as compact cars. Males and females alike are front-loaded with muscle and will attack when threatened, raising their tails to warn of a charge. Despite their bulk, they can speed across the prairie at thirty-five miles per hour—faster than most galloping horses. Their immense heads are equipped with short, curved, lethally pointed horns, and even today, more visitors to Yellowstone National Park are injured by bison than by bears.

Heavy Runner's ancestors knew better than to tackle a bison herd alone. In the summer, when bison congregated on the grasslands to mate and graze, hunters sometimes lured the animals into corrals and shot them with bows and arrows. Another strategy was for groups of families to set up camp at the base of a cliff like this one, then send a detachment of hunters to stealthily approach a herd. Over days or even weeks, the hunters would steer the bison up the ridge and between two parallel rows of rock cairns called drive lines. As the drive lines funneled the bison toward the cliff, the hunters used the cairns as blinds, crouching behind them and then, after the bison passed by, quietly following the herd. Just before the unwitting animals reached the cliff edge, the buffalo caller—a young, agile man, usually wearing a buffalo headdress—rushed across their path, yelling and gesturing to get the animals' attention. The hunters trailing the herd started making noise, too, and as they startled the bison into a stampede, the buffalo caller sprinted toward the cliff edge and leapt off, bison in pursuit.

If all went according to plan, the buffalo caller landed on a narrow ledge not far below the ridge, while the bison stampeded over his head and fell to the bottom of the precipice. The animals not killed on impact were dispatched by the people waiting below. Peering over the cliff, I wondered aloud if buffalo callers ever missed the ledge. "Maybe, but I wouldn't want to be the guy who did that and got talked about for it," Heavy Runner laughed. "He wouldn't get to be caller again."

The Blackfeet are one of four Blackfoot-speaking tribes and First Nations—known collectively as the Blackfoot Confederacy—whose ancestors relied on the bison herds of the North American plains. They cured bison hides for blankets and robes; they boiled bison bones in hide-lined pits in order to extract the fat; they combined the fat with dried bison meat and berries to make pemmican, a protein-rich trail mix. Bison were so important to human survival that they became central to human culture, too, and are still recalled in ceremonies. A few years ago, when the Brooklyn Museum commissioned a hand-painted tipi from Heavy Runner, he chose a design bequeathed to him by a mentor: a stylized bison skull with blood dripping from its jaw, and horns resembling upraised human arms.

This *pishkun*, one of many scattered across the plains, was active for hundreds of years—primarily between the 900s and the early 1500s—and it was a busy place. Excavations at the foot of the jump have uncovered compacted layers of bison bones thirteen feet deep, deposits so dense that until their archaeological importance was recognized, farmers mined them for bone-meal fertilizer.

Long before European conquest, humans and bison were shaping the plains—and each other. Bison numbers rose and fell as hunting strategies evolved, new technologies emerged, and economies changed, but even the deep kettle of blood never threatened to drain the plains of bison. In fact, Native Americans' use of fire to clear forests and expand grasslands probably boosted bison numbers, and some historians believe that European settlers encountered the bison population at its peak.

In the early 1700s, North America was home to an estimated twenty to thirty million bison, more than enough to circle the equator if laid nose to tail. The plains bison ranged from northern Mexico to southern Canada, and from Oregon to the Appalachians. In the fall of 1770, George Washington and a few companions shot five bison in West Virginia. In 1806, when explorers Meriwether Lewis and William Clark encountered a "moving multitude" of bison at the mouth of the White River in the Dakotas, the herd was so dense that it darkened the land.

While even the deadliest buffalo jumps had little effect on these masses, guns and horses left a much deeper mark. The efficiency of equestrian hunting led Native American societies to depend even more heavily on bison, and by the late 1700s, their hunters had begun to take a toll on the herds. In the 1800s, when demand for buffalo robes and leather on the East Coast and in Europe prompted commercial hunters—first mostly Native American, then mostly white—to start killing bison wholesale in both the United States and Canada, the decline accelerated. After the first transcontinental railroad was completed in 1869, the decline turned into a freefall, and by 1872, hide hunters were slaughtering more than a million bison each year, often leaving the carcasses to rot. Practiced hunters could kill as many as a hundred bison in a day, and showman Buffalo Bill Cody claimed to have shot more than four thousand in an eighteen-month spree.

When the United States entered the final decades of its war on the plains tribes, the ongoing decimation of the bison began to look like a convenient way to control the enemy. Columbus Delano, President Grant's Secretary of the Interior, believed that the disappearance of the bison would "confine the Indians to smaller areas, and compel them to abandon their nomadic customs." In 1874, when Congress passed a bill restricting bison hunting, President Grant vetoed it. Twelve years later, when Hornaday and his companions were hunting the bison that would one day be displayed in the National Museum, Hornaday estimated that there were fewer than three hundred free-roaming plains bison left in the entire United States. In Canada, there were thought to be none at all.

In the course of a few decades, the members of the Blackfoot Confederacy had lost their primary source of protein, warmth, and cultural strength. The reversal of fortune was cataclysmic, and researchers have found that its effects persist; the tribes and First Nations of the North American plains, once among the richest societies in the world, are today among the poorest. Heavy Runner, who is in his early sixties, remembers that when he was growing up on the Blackfeet reservation, his father often pointed out the hills and cliffs once used for

"After the Chase" by Montana photographer Laton Alan Huffman, taken in the late 1870s or early 1880s. Hornaday included an engraving based on this photograph in his 1889 report "The Extermination of the American Bison."

buffalo jumps. But the buffalo themselves survived only in the stories his grandparents remembered from childhood. Heavy Runner himself didn't get close to a living bison until he left the reservation for college and took a trip to Yellowstone. And though First Peoples Buffalo Jump

State Park is a place where the distant past can seem very close, the only bison in the park today stand silently inside the visitor center, stuffed and mounted.

The subspecies that scientists call *Bison bison bison* is known in the Blackfoot language as *iinii*, in Arapaho as *bii* or *heneecee*, and in any indigenous plains language as a synonym for life itself. For most of the plains dwellers who experienced the near-extermination of the bison, it was a crushing tragedy and a mortal threat. But on the urbanizing coasts of North America, in the centers of population and political power, it was at first little more than a rumor—a tale from the very edges of the known world and the very limits of the imagination.

For many who heard about the bison slaughter from afar, extinction itself was a strange and relatively new idea. Though European explorers had already extinguished dozens of bird species on islands in the Atlantic, Pacific, and Indian oceans, the assumption that species were static and enduring was not easily dislodged. Only at the very end of the previous century had studies of elephant skulls and mammoth and mastodon fossils by the French naturalist Georges Cuvier demonstrated that species were not eternal, but could and did go extinct. In 1787, Thomas Jefferson had used the Linnaean concept of a celestially centralized natural economy to argue that mastodons still existed. "Such is the economy of nature," he wrote, "that no instance can be produced of her having permitted any one race of her animals to become extinct; of her having formed any link in her great work so weak as to be broken." Before Jefferson dispatched Lewis and Clark on their famous journey across North America in 1803, he instructed them to keep an eye out for "rare or extinct" animals—perhaps holding out hope for a mastodon.

Charles Darwin, in 1859, acknowledged that humans themselves could cause extinctions, but maintained that the balance of nature's economy remained more or less constant. "We need not marvel at extinction," he wrote, for extinct species would soon be replaced by new ones. Not all shared his equanimity; a few years earlier, on the Atlantic coast of the United States, Henry David Thoreau had mourned the dis-appearance of "the nobler animals"—including cougars, panthers, lynx, wolves, beavers, and turkeys—from the countryside. In his influential 1864 book *Man and Nature*, the U.S. diplomat and polyglot George Perkins Marsh warned that human-caused extinctions were underway. "The myriad forms of animal and vegetable life, which covered the earth when man first entered upon the theatre . . . have been, through his action, greatly changed in numerical proportion, sometimes much modified in form and product, and sometimes entirely extirpated," Marsh wrote. The British ornithologist Alfred Newton, a contemporary of Darwin's, was one of the first naturalists to recognize that human-caused extinctions—what he called "the exterminating process"—could bring an entire lineage to an abrupt and permanent end.

In Africa, wealthy British colonists and travelers were learning that the animals they so enthusiastically hunted for sport and profit had their limits: the quagga, a subspecies of the plains zebra, went extinct outside captivity in the 1870s, and elephants were in alarmingly speedy decline. In the British Isles, bird lovers were beginning to realize that familiar species could be as vulnerable as those on distant Pacific islands. But even some naturalists found it difficult to believe—or, perhaps, easier to disbelieve—that humans could eliminate a seemingly plentiful species. "I believe it may be affirmed with confidence," Darwin's friend and defender T. H. Huxley opined in 1883, "that in relation to our present modes of fishing, a number of the most important sea fisheries, such as the cod fishery, the herring fishery, and the mackerel fishery, are inexhaustible." Many in North America and Europe maintained that the great herds of *Bison bison* were inexhaustible, too.

As successive reports from the North American frontier made it clear that the bison was, in fact, about to be extinguished, some saw the fate of the bison as a sad but inevitable footnote to the heroic tale of national expansion. "The bison has had his day, and now he must make room for some of his remote relations who have come to take his place," the *San Francisco Chronicle* editorialized. "And why should this not be so? The best and most valuable crop that the United States can raise is men and women." Animal welfare societies, many of which were led by women, expressed horror at the "sickening scenes of slaughter" on the plains. But for most people in the urban United States, bison were an abstraction, and the most common response to their reported plight was a shrug.

William Hornaday's decision to champion the bison was unusual both for his time and for someone of his background and experience. Orphaned as a teenager, he grew up in Indiana and Iowa on what was then the nation's frontier, often taking rambles through the woods and fields. "As soon as I became old enough to have as many thoughts as a shepherd dog, I began to appreciate the beauties and wonders of the wild creatures of the field and forest," he recalled years later. He studied taxidermy in college, and then, with typical impatience, dropped out after his sophomore year to chase fierce beasts and personal glory. He killed elephants and tigers in India and sloths and anteaters in South America, rarely if ever wondering about the longevity of their species. Before he took up the cause of the bison, he had spent much of his professional life up to his elbows in animal innards.

He was less inspired by Thoreau than by popular self-help texts such as *Getting on in the World* by William Mathews, which counseled honesty, punctuality, self-discipline, and focus in pursuit of one's chosen aim. "Having found out what you have to do," Mathews wrote, "do it with all your might, because it is your duty, your enjoyment, or the very necessity of your being." Though Hornaday moved in scientific circles, he was not an academic by training or temperament. He was, at heart, a brash frontier kid—his first employer, New York specimen dealer Henry Ward, called him "my western man"—and he was getting on with all his might.

When Hornaday arrived at the National Museum in 1882, he was surprised to discover that the institution had only a few moth-eaten specimens of *Bison bison*. He wrote to contacts all over the Great Plains about the prospects for acquiring more, and from their replies he learned, with rising alarm, that there were few if any free-roaming bison left alive in the western United States. "I received a severe shock, as if by a blow on the head from a well-directed mallet," he later wrote. "I awoke, dazed and stunned, to a sudden realization of the fact that the buffalo-hide hunters of the United States had practically finished their work." (British prime minister Lord Salisbury suffered a roughly contemporaneous revelation about the large mammals of Africa when, as the *Saturday Review* recounted, he "suddenly awoke to the fact that a grand national possession was being ruthlessly squandered, the great game of the Empire.")

It's not clear exactly what delivered the blow to Hornaday's head. Perhaps he was scandalized by the gore and waste reported by his contacts on the plains. Perhaps the prospect of such a notable extinction in his own country offended his national chauvinism—or his already well-established sense of racial superiority. Perhaps the bison's plight recalled him to his adolescence, and the solace he had found watching wildlife on the open prairie. Whatever the source, the shock was genuine, and in the spring of 1886 he bid a wistful farewell to his wife and young daughter and set off for Montana Territory. Along with two companions, he provided himself with a Remington double-barreled shotgun, a Smith and Wesson revolver, two rifles, and some one thousand cartridges. He was determined to preserve at least a few of the remaining representatives of *Bison bison* behind glass—for the sake of science, the public, and ultimately the species itself.

When the Northern Pacific Railroad deposited Hornaday and his companions in Miles City, a boozy crossroads in the southeastern corner of the territory, their welcome was friendly but dispiriting. "All inquiries elicited the same reply," Hornaday recalled. " 'There are no buffalo any more, and you can't get any anywhere.' " The hunters set out anyway, into open country strewn with bison bones and sun-seared bison flesh. They trudged through extreme heat, driving hailstorms, and slick, boot-sucking mud, enduring day after day of disappointment.

Finally, during their second expedition and seventh week on the trail, Hornaday and three of his companions came upon a small herd and shot four bison, including a cow and a young bull. In the days that followed, the party killed sixteen more animals. Hornaday and an assistant butchered the bodies and cleaned the skins, sometimes by firelight. "It is a fearful job to wash the blood out of a skin, a long, cold, tiresome job, freezing to the hands, breaking to the back," Hornaday wrote in his journal.

Just before turning toward home, in early December of 1886, Hornaday spotted a gigantic bull bison. He brought down the bull with a shot to the shoulder, but as he approached on horseback, the animal struggled upright and ran off. When Hornaday caught up after a short pursuit, he realized that the bull was "a perfect monster," the largest anyone in the party had ever seen. "In his majestic presence," Hornaday later wrote, "the finest of all our other buffalo bulls were quite forgotten, and I thought to myself: 'Until this moment I have never had an adequate conception of the great American bison.'"

The bull would become the central figure in Hornaday's museum display, and years later, the U.S. Treasury Department would use him as the model for the fierce bison that once enlivened the front of the ten-dollar bill. But the former trophy hunter professed no pride in his kill. Hornaday recalled that as he stood on the snowy prairie, pinned by the bull's dying glare, he experienced "the greatest reluctance I ever felt about taking the life of an animal."

Hornaday revealed his bison display to the public in the spring of 1888, and it was every bit the sensation he had promised. "A little bit of Montana—a small square patch from the wildest part of the wild West—has been transferred to the National Museum," gushed the *Washington Star*. "It is so little that Montana will never miss it, but enough to enable one who has the faintest glimmer of imagination to see it all for himself—the hummocky prairie, the buffalo-grass, the sage-brush, and the buffalo."

Hornaday had, as he put it, brought "the charms of wild nature within daily reach of the cribbed and confined millions," and represented the bison's predicament in new and memorable dimensions. Already, though, he wanted more. He began to dream of a national zoo where living bison could not only be displayed to the public, but bred for eventual release on the plains. For him, it was the obvious next step in what had become a righteous crusade. "It now seems necessary for us to assume the responsibility of forming and preserving a herd of live buffaloes which may, in a small measure, atone for the national disgrace that attaches to the heartless and senseless extermination of the species in the wild state," he wrote.

The Secretary of the Smithsonian backed the proposal, as did Congress, and Hornaday oversaw the purchase of the site that would become the National Zoo. But Hornaday, mercurial and chronically opinionated, fell out with the Smithsonian leadership over the zoo's final design. Though the project proceeded, Hornaday stepped down in 1890, and retreated with his wife, Josephine, and their daughter, Helen, to Buffalo, New York, where he speculated in real estate, worked on a novel, and licked his professional wounds.

Meanwhile, the number of bison continued to plummet. Though Yellowstone National Park had been created in 1872, it was intended to protect scenery, not wildlife, and bison hunting within its borders continued almost unabated. In 1897, the last four free-roaming plains bison outside the park were shot in a mountain valley in Colorado. Five years later, despite efforts by the U.S. military to control poaching in Yellowstone, the park's herd had been reduced to fewer than two dozen. Though some plains bison survived on ranches, and an isolated population of wood bison would be discovered decades later in northern Canada, the species that had traveled the continent for millennia was effectively extinct.

In 1896, after a half dozen years in Buffalo, Hornaday was hired as the director of the new zoological park that would become the Bronx Zoo. He was delighted to be out of exile, and shortly before Josephine and Helen joined him in New York City, he sent an affectionate report to the "Empress of Buffalo," as he teasingly addressed his wife. "The day is fine," he wrote, "and if you were only here, we would have a glorious day together, eating lotuses in Bronx Park."

Hornaday saw no glory in the surrounding city, however. In the decades since the Civil War, the United States had metamorphosed from a largely rural society into a largely industrial one. The railroad system had blossomed, the number of factories had more than tripled, and the country's already speedy urban growth had been accelerated by the arrival of some ten million immigrants, including an increasing number from eastern and southern Europe. New York, Hornaday observed nastily, had become an "alien city," overrun by "Jews from the slums of Riga."

Among affluent city dwellers, these changes inspired nostalgia for open spaces—and worry that without the renewing effects of wide horizons, white American masculinity would rot away in what philosopher and psychologist William James, in 1910, caricatured as "a world of clerks and teachers, of co-education and zo-ophily, of 'consumer's leagues' and 'associated charities,' of industrialism unlimited, and feminism unabashed." Such fears were bolstered by the frequent diagnosis of neurasthenia, a condition that might now be called anxiety or depression, among the well-to-do. The frontier, so recently perceived as an obstacle to modern civilization, became a standard prescription for its ills. Ailing young men were sent to the territories on hunting holidays, much as their counterparts in Britain were sent to the African colonies for a bracing dose of adventure.

For Hornaday and his closest colleagues at the New York Zoological Society, these concerns were explicitly and emphatically tied to racial anxieties. Madison Grant, who had helped found the Zoological Society when he was just twenty-nine, was a Manhattan bon vivant and one of the most effective conservationists of his era, successfully championing several state and federal laws that restricted commercial and "unsportsmanlike" hunting. A hunter himself, Grant subscribed to the common belief that hunting was an elevating pastime for the wealthy and white, but he deplored the effects of subsistence and market hunting on the numbers and genetic stock of his prized quarries.

Grant's implacable belief in the power of inheritance convinced him that his own Linnaean "subspecies"—which he called the "Nordic race," limited to people of northern European extraction—was similarly threatened by immigration and intermarriage. In 1916, he would elaborate on these ideas in a putatively scientific tome called *The Passing of the Great Race*. The book was warmly received in the United States, and, after it was translated into German, Adolf Hitler praised it as "my bible" and incorporated its claims into *Mein Kampf*. (The influence of Grant's book persists; in 2011, Norwegian extremist Anders Breivik quoted it in the rambling manifesto he wrote before murdering seventy-seven people, including sixty-nine young members of the Norwegian Labour Party.)

Another major figure in the Zoological Society was Henry Fairfield Osborn, a well-known paleontologist at the American Museum of Natural History in New York. Osborn was notoriously arrogant and skimpy with credit—he often put his own name to papers written by his subordinates at the museum—but he eagerly took up his pen to write a preface to *The Passing of the Great Race*, praising its insight and calling for the "conservation of that race which has given us the true spirit of Americanism."

Hornaday, who with Grant and Osborn formed the society's de facto ruling triumvirate, tolerated and in many cases actively endorsed these views. He, too, looked down on commercial and subsistence hunters, and habitually blamed the worst effects of hunting on people of other races, ethnicities, and nationalities. He insisted that Native Americans and white market hunters were equally responsible for the bison slaughter, despite contemporary evidence to the contrary.

Hornaday's callousness toward his fellow humans was most appallingly displayed in 1906, when he and his Zoological Society colleagues coerced a young man from the Congo, Ota Benga, into living alongside an orangutan in the Primate House. The diminutive Benga, who had been taken from his homeland by a South Carolina missionary named Samuel Phillips Verner, was said to be twenty-three but appeared far younger to North American eyes. Tens of thousands of people came to gawk as he spent his afternoons behind iron bars, weaving twine mats and returning visitors' curious stares. When a group of prominent African American clergymen protested Benga's captivity, Hornaday dismissed their concerns, claiming that the "exhibition" had scientific value. But as the swelling crowds began to harass Benga, and Benga himself started to resist his captors, public criticism grew. Finally, after three weeks of controversy, minister James H. Gordon secured Benga's release and helped him establish a life in the United States, eventually arranging for him to settle in Virginia.

In a letter to New York's mayor, Hornaday was unrepentant. "When the history of the Zoological Park is written," he predicted, "this incident will form its most amusing passage." Bluster, however, could not heal the wounds inflicted. Ten years later, after concluding that a return to the Congo was impossible, Ota Benga shot himself through the heart.

During the months that Hornaday worked on his bison display at the National Museum, one of the few visitors to circumvent the privacy screens was the young Theodore Roosevelt, who, when he stopped by in 1887, was fresh from a failed campaign for mayor of New York City. Roosevelt, a devoted hunter, credited what he would famously call the "strenuous life" with his recovery from a sickly boyhood. He shared Hornaday's fascination with bison, and he had recently founded the Boone and Crockett Club, a group of wealthy hunters eager to maintain free-roaming populations of the animals who made their pastime possible. The club, whose leadership included Roosevelt's friends Grant and Osborn, would incubate and support the Zoological Society and become a hub of the early conservation movement in the United States.

Roosevelt's dedication to conservation, as both public servant and private citizen, is legendary. As president of the United States from

1901 to 1909, he designated almost a quarter of a million acres of public land as national parks, forests, monuments, and wildlife reserves.
Like Grant, Osborn, and Hornaday, his commitment to saving the
bison was both genuine and infused with racism, for he believed that
bison were essential to the pursuit of the strenuous life—which, in
turn, was essential to the survival of white masculinity. The struggle
to advance the frontier, Roosevelt wrote, had produced an "intensely
American stock"—a stock tested and bettered by "conflicts with the
weaker race." Born too late to fully test his mettle on the North American frontier, Roosevelt would turn to frontiers overseas; he rose to
national renown by fighting in Cuba during the Spanish–American
War, and as president he sought to expand the influence of the United
States abroad.

Roosevelt did not forget Hornaday and his taxidermied herd, and in
1905 he and several fellow Boone and Crockett members helped Hornaday organize the American Bison Society, a private organization dedicated to bison restoration. The society enlisted financial support from
urban elites, often presenting bison as an answer to intertwined anxieties about industrialization, masculinity, and perceived racial decline.

Hornaday, prodded by his fellow bison advocates, had meanwhile
restarted his stalled plans to raise bison in captivity, beginning with
seven animals purchased from ranchers in Texas and Oklahoma. It
wasn't easy; several early captives, unused to the lush green grass on
the zoo grounds, died after overindulging, and breeding proceeded
slowly. But bison had enough in common with domesticated cattle that
they responded to familiar husbandry techniques, and after a few years
of trial and error, Hornaday established a small bison herd in the heart
of the Bronx.

Hornaday didn't know if his bison could survive without constant
care, much less successfully reproduce, but he believed they were the
best remaining hope for *Bison bison* and he was determined to move
them out of the city. After New York's governor vetoed a plan to release
Bronx-raised bison in the Adirondack Mountains, Hornaday focused

his energies on Oklahoma Territory, where Roosevelt had created the Wichita National Forest and Game Preserve in 1905. The preserve's rich grasslands, perfect for a herd of bison, had been seized from the Apache, Comanche, and Kiowa.

Which brings us to the central irony in Hornaday's irony-soaked story: for Hornaday and his allies, the rescue of the bison had nothing to do with the people who had depended on the species—and a great deal to do with their own illusions about themselves. Protecting the American bison, to them, meant protecting a perniciously exclusive version of national progress.

Racial anxieties were not the only fuel for Hornaday's conservation crusade. The swift and total extinction of the passenger pigeon, last confirmed outside captivity in 1900, had demonstrated that commercial hunters could destroy even the most abundant species. ("There will always be pigeons in books and in museums," conservationist Aldo Leopold would reflect four decades later, "but these are effigies and images, dead to all hardships and to all delights.") Millions of people had seen or heard of Hornaday's bison display at the National Museum, and the extinction of the bison, once seen by so many as an inevitable if regrettable consequence of expansion, had come to be viewed as a needless tragedy. "It is now generally recognized as ethically wrong to jeopardize the existence of any animal species," the California naturalist Joseph Grinnell—a distant cousin of conservationist and Boone and Crockett Club member George Bird Grinnell—declared a few years later. Joseph Grinnell may have been overly optimistic, but he was essentially right: public attitudes toward extinction had changed for good.

More cynically, opposition to the bison's extinction was no longer a political risk. By the early 1900s, the species was so far gone that its protection posed no threat to existing financial interests or military strategies, and the return of a few bison to the prairie carried no significant costs, monetary or otherwise.

On October 11, 1907, fifteen city-bred bison were coaxed into wooden crates at the Bronx Zoo, then loaded on to railcars at New York's Fordham Station. Over the next seven days, accompanied by a clerk from the Zoological Society and an Oklahoma cowboy and entertainer named Frank Rush, they traveled nearly two thousand miles to their new home, making a stop at the state fair before arriving in Cache, Oklahoma. There they were greeted by an enthusiastic crowd that included the Comanche chief Quanah Parker. Still in their crates, the restless bison were loaded into wagons and transported the remaining twelve miles to the preserve, where they were swabbed with crude oil to ward off ticks. Then they were released into a pen to await the completion of a fence around their eight-thousand-acre enclosure.

The Wichita bison would not wander like their ancestors, but they would roam further on the prairie than almost any bison had for decades. Their arrival was recognized nationwide as cause for celebration, and Hornaday, while far from the only champion of the bison, was applauded as their savior. "Director Hornaday of the Bronx Zoological Park deserves the gratitude and encouragement of the Nation as the chief preserver from extinction of the American bison," editorialized the *New York Times*. When the Oklahoma herd welcomed a bull calf in November, caretakers named him Hornaday.

The Wichita herd flourished, and in 1908, Hornaday convinced Congress to designate another bison preserve, this time on nineteen thousand acres that the federal government had recently carved out of the Flathead reservation in northwestern Montana. The American Bison Society raised funds to buy thirty-four bison for the newly christened National Bison Range, contributing to a starter herd that by the early 1920s had grown to five hundred. The Northern Pacific, which had transported Hornaday, his companions, and their personal arsenal across the Midwest to Montana, began to serve meat from the National Bison Range in its dining cars, assuring its customers that the cuts offered a particularly patriotic flavor of conservation.

When the American Bison Society finally stopped collecting dues in the mid-1930s, there were twenty thousand bison in North America, most in large, enclosed reserves. The herds were nowhere near the "moving multitudes" that once populated the plains, but they were enough to ensure the persistence of the species for a good long while.

Though Hornaday restored *Bison bison* to the United States, he didn't restore the species to its place on the landscape. Only in 1866 had zoologist Ernst Haeckel coined the word "ecology," or *Oecologie*, to describe the study of the relationships among organisms and between organisms and their environment—or, as he put it, "the study of all those complex interrelations referred to by Darwin as the conditions of the struggle for existence." (The words "ecology" and "economy" are both rooted in *oikos*, the ancient Greek term for family and household.) When Hornaday shipped bison from the Bronx to the Oklahoma grasslands, ecology had barely begun to develop into a formal discipline, and Hornaday was not an ecologist. He treated bison essentially as livestock, letting them loose without much consideration of their role in the living network Darwin had described as the "entangled bank."

But, as future generations of scientists and conservationists would learn, bison had been part of the prairie in a very real sense. They fertilized the soil with their dung, flattened saplings and shrubs with their hooves, and grazed so intensely each spring that the native grasses extended their growing season. In Yellowstone, bison still have more influence on the timing of plant growth than the weather does.

Over the past century, the expansion of industrial farming and the continued absence of large herds of free-roaming bison have shrunk the total extent of the North American tallgrass prairie from the size of Texas to little bigger than Delaware, turning it into one of the most endangered landscapes in the world. The northern shortgrass prairie, where Hornaday once searched for bison, is similarly diminished.

The prairie that remains is fundamentally different. Canadian ecologist Wes Olson, who studies what he calls the "bison snot ecosystem" and the "bison patty ecosystem," found that as bison snuffle along the prairie, the microbes they take in through their noses and mouths help break down the cellulose in the grass they eat. Each of the result-ing soft, wet piles of dung can support more than a hundred different species of insects. When bison were plentiful, these insects fed a com-munity of birds and small mammals, but many of those species are now seldom seen on the prairie. Without bison—without bison snot, bison crap, and everything in between—the prairie is a smaller and quieter place.

In the 1990s, the Blackfeet Nation purchased about a hundred bison from the Hornaday reserves. Ervin Carlson, a Blackfeet rancher and the head of the tribal agricultural department, oversaw the herd, and as it grew he began to wonder if truly free-roaming bison could be restored to the Northern Rockies. In 2008, he started to talk with Keith Aune, a biologist who had recently retired from a long career with Montana's state wildlife agency, about the possibility of returning more bison to the Blackfeet reservation.

Aune shared Carlson's deep interest in bison. A few years earlier, he had paid a visit to the Bronx Zoo—which also serves as the headquar-ters of the international Wildlife Conservation Society—and proposed to help reopen the doors of the American Bison Society. Through the revived bison society, he hoped to reconnect the plains bison with its landscape and, crucially, with the people who live on that landscape today.

Over the next several years, Aune secured funding and support for bison restoration on the Blackfeet reservation from the Wildlife Con-servation Society, and members of the Blackfoot Confederacy founded a group called the Iinii Initiative, after the Blackfoot word for bison. The Iinii Initiative negotiated an agreement called the Buffalo Treaty, which dedicates its signatories—more than thirty tribes and First Nations from the United States and Canada, some of whom have a long history of disagreements with one another—to the active pursuit of bison restoration.

On a windy, gray afternoon in April 2016, members of the Blackfoot

Confederacy gathered on the Blackfeet reservation, on the flat southern bank of the Two Medicine River. Elders wrapped in the bright woolen stripes of Hudson's Bay blankets and coats stood beside fidgeting elementary school kids, and everyone watched the bluff above the river. A line of young horseback riders appeared, silhouetted against the sky. Behind them rattled a caravan of white livestock trailers, loaded with bison calves from Elk Island National Park in Alberta. The calves were direct descendants of some of the last bison in Blackfeet territory, a half-wild herd purchased by the Canadian government in the early 1900s from Michel Pablo, a man of Blackfeet ancestry. There were eighty-eight bison in all, and after more than a century, they were returning to the home of their forebears.

• • • •

Blackfeet bison manager Sheldon Carlson (Ervin's first cousin), standing beside a trailer holding bison calves from Elk Island National Park, April 2016.

While the return of the Elk Island bison was heartily celebrated, on the reservation and off, it didn't happen without controversy. Cattle ranchers, including some tribal members, worry that large bison herds—fenced in or not—will compete with them for grazing land and market share. Bison in and around Yellowstone National Park also carry the infectious disease brucellosis, which affects cattle. Direct transmission of brucellosis from bison to cattle has never been documented, but the possibility has nonetheless created great apprehension, and further poisoned the relationship between ranchers and bison advocates. (Bowing to public concerns about brucellosis, Yellowstone managers agreed to limit the size of the park's herd, and as a consequence several hundred Yellowstone bison are shipped to slaughter or shot under permit each year.)

There is no doubt that bison can be inconvenient: colossal, dangerous when provoked, and oblivious to traffic and property lines. In Gardiner, Montana, near the northern entrance to Yellowstone, it's not unusual to see bison grazing on the high school football field, and students serve out detention by clearing bison patties off the turf.

Bison restoration, despite its ties to the past, can also represent unwelcome change. On forty-four thousand acres of grassland in central Montana, a group called the American Prairie Reserve has reestablished a herd of more than eight hundred bison, many of them descendants of the animals Hornaday raised at the Bronx Zoo. Reserve founder Sean Gerrity, a Montana native and successful Silicon Valley entrepreneur, hopes to one day buy or lease more than three million acres of public and private lands, allowing the bison to resume something like their former movement patterns. The reserve works only with willing landowners, and has brought jobs and other benefits to a region long battered by drought and economic shocks. Still, many cattle ranchers see the effort as an attack on their livelihoods—and their identity.

Even among those opposed to bison restoration, however, there is a sense that bison are, as biologist Lee Jones puts it, "an important something." Their importance to history, culture, and the North American landscape is widely appreciated. Since the revival of the American Bison Society, a diffuse, largely leaderless movement of tribes and First Nations, private conservation groups, state and federal agencies, and others has built support for bison restoration, and its projects range from northern Mexico to the Arctic Circle.

In Europe, the wisent—which, like *Bison bison*, is a descendant of the prehistoric steppe bison—is making a similar comeback. Long restricted to a single population on the border of Poland and Belarus, wisents have been reintroduced in Germany, France, the Netherlands, and elsewhere, settling into landscapes that lost their ancestors centuries ago.

The rolling hills of the Blackfeet reservation were covered with bright-yellow balsamroot flowers, and, in the distance, the ridgeline of the Rockies was frosted with thinning snow. It was early June, and in Browning, the reservation's largest town, clouds of dust were beginning to rise from the unpaved back streets. In the gas stations and cafes, talk dwelled on the past winter: who ran out of propane, who lost the most cattle, who was in the worst fix.

On the grassy banks of the Two Medicine, though, the bison were doing fine. The Blackfeet Nation manages two bison herds totaling about eight hundred animals, with one designated for local use and the other—the Elk Island herd—for conservation. The Elk Island bison were fully grown, and a new crop of calves stood knock-kneed next to their mothers in a sunny paddock, their short, golden-brown hair contrasting with the adults' heavy, half-shed winter coats. As Carlson and I got out of his truck and approached, the animals snorted warily. One calf lifted a tail and peed abundantly, out of fear or defiance or both. I recalled a comment by Blackfeet elder Charlie Crow Chief: "They recognize a stranger." Carlson chuckled, surveying them proudly. "The way I look at this herd is, they were in storage. They put themselves in storage until they were ready, and we were ready."

The bison, feeling the change in the weather, paced the fenceline and tussled with one another. While there are an estimated half a million bison on the North American plains, some thirty thousand managed primarily for conservation rather than commercial sale, politics has kept the fences in place, and only a very few are fully free-ranging. Carlson, now the manager of the Blackfeet's Buffalo Restoration Project, looks forward to a day when bison roam the prairie like elk or deer, and he may get to see it; if ongoing discussions succeed, the Elk Island herd will be permitted to move north at will, entering Glacier National Park and perhaps even crossing the international border.

When Hornaday rescued the bison from extinction in the United States, said Carlson, he wasn't thinking about the Blackfeet, but the two are inextricable. "Back then, some people thought that if you killed the bison, you killed the Indian too," he said. "Well, the buffalo are here today, and so are we."

The Elk Island bison are left largely to themselves, and most tribal members see them only from a distance. The animals haven't restored the prairie, or the fortunes lost when their ancestors were slaughtered. But their presence is felt all over the reservation, and has inspired projects ranging from a bison-meat lunch program at the Blackfoot language immersion school to a reenactment of a buffalo jump at the local high school (minus the real bison and the mortal danger). Students take part in the annual harvest of the local-use herd, spattering themselves with blood as they learn to butcher a bison. Not long ago, Lyle Heavy Runner and a friend leased a few acres of land adjoining First Peoples Buffalo Jump State Park and raised a herd of Spanish barbs, the hardy breed of horses that plains tribes once used to hunt bison. Heavy Runner donated the horses to the Blackfeet bison program, which has used them to move its herds.

Bison aren't as fundamental to human survival as they were in the days when buffalo callers leaped off cliff edges. Then again, maybe they are. During a meeting in the Blackfeet Nation council chambers, bison project staff and visiting conservationists listened as elders spoke about the revival of the bison. Their generation was born decades after the Blackfeet lost their bison, and came of age decades before the Elk Island herd returned, but their reverence for the species was clearly profound. Through tears, Betty Crow Chief said that her brother, while hospitalized after a severe heart attack, had seen images of bison move across the wall of his room. Many listeners nodded, unsurprised.

The bison display that Hornaday built for the Smithsonian remained part of its museum exhibits for more than seventy years. In 1957, when Smithsonian staff dismantled it, they found a small metal box in the base of the display, beneath the layer of Montana dirt. Inside was a copy of a *Cosmopolitan* magazine article Hornaday had written about his expeditions to Montana Territory, inscribed with a note from the man himself. With his usual combination of arrogance, insight, and touchiness, he had anticipated that his work would endure—and that future generations would find it lacking:

> *To my illustrious successor: Dear Sir, Enclosed please find a brief and truthful account of the capture of the specimens which compose this group. The Old Bull, the young cow, and the yearling Calf you find here were killed by yours truly. When I am dust and ashes, I beg you to protect these specimens from deterioration and destruction. Of course they are crude productions in comparison with what you may produce, but you must remember at this time (A.D. 1888. March 7.), the American School of Taxidermy has only just been recognized. Therefore give the devil his due and revile not.*
> *Wm. T. Hornaday. Chief Taxidermist, U.S. National Museum.*

The Smithsonian staff had indeed intended to destroy the display, but Hornaday's plea gave them pause. They donated the six figures to various Montana collections, which over time stashed them in storage units across the state. In the 1980s, an Oregon naturalist named Douglas Coffman began a search for the figures, eventually reuniting them at the Museum of the Northern Great Plains in Fort Benton, Montana, a Missouri River town whose tourist trade subsists on frontier nostalgia. The display needed some minor repairs—one of the figures got a new nose—but Hornaday had protected his art from time and insects by treating it with arsenic.

Today, the figures stand in the heart of the small museum, gathered on a simulacrum of Montana prairie much like the one Hornaday built behind his privacy screens. They are reverently illuminated and impressively massive, and the illusion of life still startles. They look taller and heavier than the bison on the plains today, many of which carry domestic cattle genes from crossbreeding. The glass eyes of the largest figure, the bull Hornaday faced down and shot with such regret, follow visitors around the room, and it's easy to imagine that the muscular hump above its shoulders is tensed for action. Sandy, the buffalo calf that Hornaday and his men brought back alive from their first expedition—and displayed, briefly, on the museum lawn until he died from eating too much clover—is memorialized here, too, his remains posing warily in the shadow of the much larger adults.

These figures don't snort or bellow, or fertilize the prairie with dung. They don't paw the ground or toss their heads in threat, and while you can get far closer to them than any human should get to a bison, they ultimately don't tell you much about who they used to be. As embodiments of Hornaday's uneasy legacy, I'll take the living bison calves stretching their gangly legs on the Blackfeet reservation. But it's not hard to understand why so many people lined up to see these deft arrangements of skin and clay—and how each encounter between human and hide fed the campaign to revive the bison in the United States. When I saw them, posed as if about to step off their tiny patch of prairie, I couldn't help but want them to run.

• • • •

For the conservation movement, the American bison was a defining cause, uniting ambitious, patrician white men whose admiration for the bison—and other large mammals, at home and abroad—was rooted in nationalism, racism, and their own love of the hunt. They formed what historian Stephen Fox calls an "elitist conspiracy," dedicated to their trophies and themselves.

For Hornaday, the rescue of the bison was one chapter in a very long life, and he rarely ran in place. In his later years, he supported conservation laws that shielded other species from the bison's fate. He convinced the U.S. government to sign the first international wildlife conservation treaty, a seal protection agreement with Russia, Britain, and Japan. He inveighed against the slaughter of birds for the millinery trade. And when he was in his seventies, he abetted an uprising within conservation itself, encouraging a New York bird lover named Rosalie Edge to confront the sportsmen who once supported Hornaday—and to insist, as Hornaday had not, that they speak up for species they didn't even like.

Chapter Three

THE HELLCAT AND
THE HAWKS

On the morning of October 29, 1929, at the southern tip of Manhattan Island, jittery bankers, stockbrokers, clerks, and messengers crowded into the great hall of the New York Stock Exchange. Millions of citizens had invested in the soaring stock market in recent years, many draining their savings to buy shares. In early September, stock prices had fallen sharply, recovered, then dropped and kept dropping. For days, clerks had been working through the night to handle the demands of panicked sellers; by the time the opening gong sounded on that Tuesday morning, the stock exchange's medical department was prescribing sedatives to anxious traders.

Five miles north, in her brownstone on East Seventy-second Street, Mrs. Charles Noel Edge put on an elegant dress and walked to nearby Central Park. It was a cloudy, unseasonably cold morning, but as soon as she stepped into the park, she lifted her binoculars and began to look for birds. She wandered west along the paths and through the patches of woods, noting a robin, a starling, a grackle, a junco, and several species of sparrows. Though she arrived late to her destination, the American Museum of Natural History, she was, as she later remembered, "happy and relaxed as one is after enjoying the companionship of birds."

When she entered the small ground-floor room where the National

Association of Audubon Societies was conducting its twenty-fifth annual meeting, the assembly stirred with curiosity. Edge had been a life member of the association for years, but annual meetings tended to be familial gatherings of directors and employees, and those present were clearly surprised by the appearance of an ordinary member, life or otherwise. As she was shown to a front-row seat, she noticed a hermit thrush perched on the window ledge outside.

Edge listened as Theodore Palmer, a zoologist and member of the board of directors, finished a speech extolling the association. She would recall the scene with regal irony: "It was my first experience hearing the Audubon Society praise itself, and it impressed me to know how great and good it was." Palmer's pride, however, was surely genuine. The national association, which represented more than a hundred local societies, was the leading conservation organization in North America—if not the world—during a time of intense public interest in wildlife in general and birds in particular. The directors of the association were widely respected scientists and successful businessmen, some of whom must have had riches at stake downtown.

As Palmer concluded his remarks, he mentioned that the association had "dignifiedly stepped aside" from responding to "A Crisis in Conservation," a recently published pamphlet that accused the National Association of Audubon Societies of serving sportsmen better than birds. Edge raised her hand and stood to speak, for "A Crisis in Conservation" was her reason for attending. "What answer can a loyal member of the society make to this pamphlet?" she asked. "What *are* the answers?"

Edge, at the time, was almost fifty-two years old. Slightly taller than average, with a stoop that she would later blame on hours of letter-writing, she favored black satin dresses and fashionably complicated (though never feathered) hats. She wore her graying hair in a simple knot at the back of her head. She was invariably well-spoken, with a plummy, cultivated accent and a habit of drawing out phrases for emphasis. She could be deceptively diffident when it served her purposes, but her sharp, strikingly pale blue eyes were constantly taking in her surroundings, and her characteristic attitude was one of imperious vigilance—as a *New Yorker* writer once put it, "somewhere between that of Queen Mary and a suspicious pointer."

Edge's questions to the Audubon directors were polite but piercing. Was the association tacitly supporting bounties on bald eagles in Alaska, as the pamphlet stated? Had it endorsed a bill that would have allowed wildlife refuges to be turned into public shooting grounds? Her inquiries, as she recalled years later, were met with leaden silence—and then, suddenly, outrage.

Frank Chapman, the museum's bird curator and the founding editor of *Bird-Lore*, the Audubon association magazine, rose from the audience to furiously condemn the pamphlet, its authors, and Edge's impertinence. Chapman's museum colleague Robert Cushman Murphy, curator of seabirds and a well-known Antarctic naturalist, stood up to argue for the lady's right to ask questions. Murphy, to emphasize his point, inquired if Edge were a member of the association, and she cheerfully assured him that she was a life member. This news placated no one, and several more Audubon directors and supporters stood to berate the pamphlet and its authors. According to the meeting minutes, Murphy himself called it "unfortunate in tone from cover to cover," and said that it contained "far more of spite than true zeal for conservation."

Edge persevered through the clamor, marshaling what she remembered as "all the friendliness possible." Did the association have no response whatsoever to the pamphlet's accusations, which to her seemed "too grave to be ignored with honor"? "I fear I stood up very often," she recalled with unconvincing remorse.

When Edge finally subsided, association president T. Gilbert Pearson—whom Edge tartly described later as a "rotund, little man"—rose from his seat on the platform at the front of the room. He informed Edge that her questions had taken up the time allotted to the showing of a new moving picture, and lunch was getting cold. Edge joined the meeting attendees for a photograph on the museum's front steps, where she managed to pose among the directors. She declined Murphy's polite invitation to lunch with the group in the Bird Hall, preferring to return to Central Park and its living birds.

By the end of the day, Edge and the Audubon directors—along with the rest of the country—would learn that stockbrokers on the exchange floor had sold a record sixteen million shares. Stock values had fallen by billions of dollars, and families rich and poor were ruined. The day would soon be known as Black Tuesday, the beginning of the Great Depression.

That morning, however, the Audubon directors were reacting to what they saw as a more immediate threat. During the commotion in the meeting room, Edge remembered, association secretary William Wharton approached her and whispered that "if the directors answered the pamphlet, it would only give it publicity, which would surely call it to the attention of Dr. Hornaday, who would take I do not know what horrid advantage of the situation."

Even as the directors maintained their silence, their fears had already been realized. William Hornaday had read "A Crisis in Conservation," and he soon heard about Edge's role in the meeting. A few days later, he sent Edge a letter, complimenting her on having "the courage of a lion." He invited her to call on him, and she accepted immediately.

Rosalie Edge and William Hornaday were opposites in many ways. While Hornaday was a "western man" who prized frontier self-sufficiency and macho heroics, Edge was born into the most exclusive social circles of Gilded Age New York. Edge was a committed feminist and, by the time she met Hornaday, a veteran of the suffrage movement; Hornaday supported voting rights for white women, but he clung to his noxious views of people of other races, classes, and nationalities. ("Toward wild life the Italian laborer is a human mongoose," he wrote. "Give him power to act, and he will quickly exterminate every wild thing that wears feathers or hair.") Edge was new to the work of protecting species; Hornaday had been steeped in the politics of conservation for decades.

Hornaday and Edge also had a great deal in common. Both had lived abroad, he as a trophy hunter and she as the adventurous young wife of a British civil engineer based in East Asia. Both had watched wildlife during times of heartbreak, he as an orphaned adolescent and she during and after the breakup of her marriage. Both could be charming, though when threatened, Hornaday ran hot while Edge ran cold. (She once severed a decades-long friendship with a single icy reprimand: "I did not like your letter.") Both Hornaday and Edge had a gift for language, exhibited in his popular books and in her publicity work for the formidable New York State Woman Suffrage Party. Both tended to exaggerate when they felt their cause was just. And both, despite their own wealth and position, delighted in rattling the power structure of the conservation movement, which they saw as too sympathetic to sportsmen and not sympathetic enough to the species that sportsmen despised. Edge, writes her biographer Dyana Furmansky, was "a citizen-scientist and militant political agitator the likes of which the conservation movement had never seen."

From their first meeting, the two got along famously. Edge approached Hornaday, then over seventy, with some diffidence, aware of both his long career and what she called "her ignorance of the whole conservation problem." While she recognized his bellicosity—"He hated his enemies, but gloried in the fact that he possessed a large collection of them, all species being represented," she wrote—she encountered his best side, finding him to be a "kindly, generous, humorous, charming, and learned old gentleman." Years later, she warmly recalled that he paid her the compliment of treating her "as a fellow man."

With Hornaday's support, Edge would continue to pursue the charges made in "A Crisis in Conservation," discomfiting not only the directors of the Audubon association but most of the conservation establishment. She would, in time, collect at least as many enemies as Hornaday, but she would glory in the fact that they were mostly human.

• • • •

"What is this love of birds? What is it all about?" Edge wondered. "Would that the psychologists might tell us." Maybe it's about the mysteries of flight and migration, or the way birds pass through our daily lives as other species do not. Or maybe it's that birds always seem especially in need of sanctuary from ourselves.

Humans have probably been extinguishing bird species for thousands of years. Fossil evidence suggests that as prehistoric Polynesians expanded across the Pacific, they drove hundreds or even thousands of island bird species into extinction through hunting, deforestation, and the introduction of rats and other unfamiliar predators. Beginning in the 1500s, European explorers and their associated predators—dogs, pigs, and more rats—continued the destruction on islands worldwide. In the 1600s, on the island of Mauritius, east of Madagascar, the dodo— a pigeon relative whose large size and ground-nesting habits had long suited it for survival—was famously obliterated in a matter of decades. Throughout the southern oceans, species of rails—shy, long-toed birds who often lose the ability to fly when isolated on islands—disappeared by the dozen, and local extinctions of more widely distributed species became routine. "Formerly, a great many flamingoes were also found about this isle; this is a large and beautiful sea fowl of a rose colour," lamented French writer Bernardin de Saint-Pierre after his army service on Mauritius in the mid-1700s. "They say also that three of them did yet remain, but I never saw one."

Since the 1500s, humans have entirely eliminated more than one hundred and fifty bird species. No other class of vertebrates has suffered more extinctions. (The only other class of animals thought to have lost more species than Aves is Gastropoda, the snails and slugs.) And those are just the extinctions we know about.

Birds were and still are hunted not only for meat, but for beauty. Aztec artisans decorated royal headdresses, robes, and tapestries with intricate featherwork designs, sourcing their materials from elaborate aviaries and from far-flung trading networks that secured toucan and parrot plumes for noble garments. (Aztec and other industrious Pre-

Columbian bird traders transported live birds for such great distances, and in such large numbers, that they are believed to have altered species distributions throughout Latin America.) In the 1500s, the shimmering featherwork images of monsters, serpents, and birds that Spanish conquistadors brought back from their subjugation of the Aztecs enchanted European elites and roused their interest in feathers as fashion. The continent's first feather craze was kicked off by Marie Antoinette in 1775, when the young queen started to decorate her towering powdered wig with immense feather headdresses. French nobles rushed to imitate their monarch: "The head-dress of our women rises higher and higher," noted a journalist at the time. The three-foot-high confections forced women to kneel in carriages and duck into theater boxes, leading one observer to dryly predict a "revolution in architecture."

The fad waned after a few seasons, but feathers continued to be used in millinery, and a century later the craze returned in earnest. This time, feathered finery was not reserved for the rich; ready-to-wear fashions and mail-order firms made it available to women of lesser means in both Europe and North America. By the 1880s, women were decorating their hats with not only individual feathers but the stuffed remains of entire birds, complete with beaks, feet, and glass eyes—sometimes accompanied by arrangements of foliage, butterflies, and taxidermied rodents.

Ornithologist Frank Chapman, who would be the first to scold Rosalie Edge for questioning the Audubon directors, reported in 1886 that of the seven hundred hats whose trimmings he observed during walks in New York City, five hundred and forty-two were decorated with feathers from forty different bird species, including bluebirds, pileated woodpeckers, hummingbirds, and egrets.

To some city dwellers, these displays were simply a welcome reminder of the rest of life. Their toll on birds, however, was enormous. In 1886, as Chapman was surveying feathers in Manhattan and Hornaday was hunting the last few bison in Montana Territory, an estimated five million North American birds were killed to supply the millinery industry. In 1912, when Hornaday sent a young ornithologist named William Beebe to investigate the feather market in London, Beebe reported that, during the previous year, the plumage of nearly half a million dead birds—mostly egrets, herons, hummingbirds, and birds of paradise—had been bought and sold by four firms alone. Plume-hunters in Florida, finding that nesting egrets were easy targets, often shot into rookeries, skinning and abandoning the bodies of adult birds and leaving the orphaned chicks to die.

Even before the craze began, European bird-lovers had begun to understand that humans could destroy other species—not just on out-of-the-way islands but nearby. The great auk, a chunky, flightless seabird first formally described by Linnaeus, was once common on rocky coast-lines throughout the North Atlantic. Like the bison, the species was long hunted by humans for food. But in the 1700s, commercial hunters in Europe and North America also started killing great auks for their feathers, which were used in bedding, and by the end of the century the species was close to extinction. The great auk was last documented in the British Isles in 1834, and ten years later, on a rocky outcrop off the coast of Iceland, a crew organized to collect birds for a specimen dealer caught and strangled what may have been the last surviving pair. British ornithologist Alfred Newton, whose brother had sent him shipments of dodo bones while serving as a colonial secretary on Mauritius, sailed to Iceland in the late 1850s in the faint hope that the great auk had not yet gone the way of the dodo. He returned disappointed. "For all practical purposes," he concluded, "we may speak of it as a thing of the past."

As the feather trade expanded and the avian death toll mounted, male conservationists on both sides of the Atlantic blamed women "and their thoughtless, stupid devotion to 'style,'" as Hornaday wrote in a *New York Times* diatribe in 1913. Hornaday called women a "scourge to bird life all over the world," and Chapman gave an address called "Woman as Bird Enemy." Virginia Woolf, in response to such rebukes, spared no sympathy for "Lady So-and-So" who decides that "a lemon-coloured egret is precisely what she wants to complete her toilet," but ended her public letter with a sharp reminder: "the birds are killed by men, starved by men, and tortured by men—not vicariously, but with their own hands."

The accusation applied beyond the plume trade. Ornithologists,

THE "EXTINCTION" OF SPECIES;
OR, THE FASHION-PLATE LADY WITHOUT MERCY AND THE EGRETS.

"The 'Extinction' of Species; Or, the Fashion-Plate Lady Without Mercy and the Egrets," 1899 illustration from the British satirical weekly *Punch*.

whose ranks at the time were overwhelmingly male, regularly collected eggs from nests and shot and skinned birds for closer study, sometimes by the thousands. Chapman, who had abandoned a promising banking career when he fell in love with ornithology, almost single-handedly guaranteed the extinction of the Carolina parakeet when he shot fifteen of them during an 1889 expedition to Florida's Atlantic coast. Like Hornaday's quest for surviving bison, Chapman's pursuit of the nation's last endemic parrots was motivated by the species' precipitous decline, and as he searched the piney banks of the Saint Sebastian River, he planned to kill only a few parakeets for posterity. But as soon as he touched the brilliant plumage—a rainbow of green, yellow, red, and orange hues—his ambitions grew. "I have met *Cornurus* and he is mine," he proclaimed after collecting his first four specimens. Once he had nine birds in hand, he resolved to stop ("far be it from me to deal the final blows," he wrote in his journal), but his pledge lasted only two days. "The parakeets tempted me and I fell; they also fell, six more of them," he reported.

Chapman's passion for possession was not unusual among his contemporaries. While many ornithologists were revolted by the plume trade, not a few worried that strict bird protection laws would interfere with their own work. In 1902, when the president-elect of the American Ornithologists' Union was invited to attend an Audubon Society meeting, his vinegary refusal was only partly in jest: "I do not protect birds. I kill them."

Virginia Woolf might have noted that while affluent women were the primary customers of the feather trade, they were also among its most effective opponents. As binoculars and other portable optics had become cheaper and more widely available in the late nineteenth century, bird enthusiasts realized that they no longer needed to kill birds to get a good look at them, and birdwatching—by foot, bicycle, and later automobile—became a fashionable pastime in both North America and Europe. Some of the most prominent early birdwatchers were women. Florence Merriam Bailey, the younger sister of the eminent zoologist C. Hart Merriam, popularized amateur birdwatching in the United States with her book *Birds Through an Opera Glass*, considered the first modern birding guide.

Enthusiasm for watching live birds soon began to compete with the enthusiasm for wearing dead ones. In 1889, on the outskirts of the industrial city of Manchester, England, a thirty-six-year-old philanthropist and bird enthusiast named Emily Williamson founded the Society for the Protection of Birds, a women's organization whose members pledged to "refrain from wearing the feathers of any bird not killed for purposes of food." They made an exception for ostrich feathers, as most of those were plucked from farmed birds—a loophole that elicited some patronizing scorn: "Not a *very* severe self-denying ordinance that, Ladies?" asked the editors of the satirical weekly *Punch*, who nonetheless supported the campaign. The society soon merged with the Fur, Fin, and Feather Folk, a group of similarly-minded London-area women, and in 1904 it received its royal charter. Today, the Royal Society for the Protection of Birds is one of the largest wildlife conservation organizations in Europe.

In the United States in 1896, Harriet Hemenway, a wealthy Bostonian from a family of abolitionists, took on the millinery industry after reading a vivid press account of the commercial egret slaughter in Florida. Hemenway and her cousin Minna Hall hosted a series of strategic tea parties for their poshest acquaintances, eventually persuading some nine hundred women to boycott feathered fashions. Realizing that their cause would benefit from the support of prominent men, Hemenway and Hall invited a select group of local businessmen and male ornithologists to help revive the Audubon movement, which had stalled shortly after its founding a decade earlier.

The movement's namesake, John James Audubon, was a man of contradictions. Born in Haiti in 1785 and raised in France, long rumored to have had African ancestry, Audubon opposed the abolition of slavery and, while living in Kentucky and Louisiana during the 1810s

and 1820s, was a slaveholder himself. Despite spells of severe depression, he used his artistic talent to express his fascination with birds, a
feeling he described as "bordering on phrenzy." His masterwork, *The
Birds of America*, is a collection of 435 life-sized, hand-colored prints
that took more than a decade to complete and cost the modern equivalent of two million dollars to produce. "My best friends," he recalled
in his journal, "solemnly regarded me as a mad man." The collection,
when it was finally completed in 1838, was a sensation on both sides of
the Atlantic, admired by critics and kings for its scientific detail and
visual poetry. (His few detractors sniffed that he should have displayed
the prints in Linnaean order.) "A magic power transported us into the
forests which for so many years this man of genius has trod," raved one
French observer. Hornaday, who encountered *The Birds of America* as
a college student, remembered the experience of turning its oversized
pages as the "discovery of a New World."

Audubon died in New York in 1851, but his legacy was carried on
by George Bird Grinnell, the conservationist and hunter who would

Left: John James Audubon in wolfskin coat, 1826.
Right: Hand-colored engraving of the "Snowy Heron, or White Egret," now
called the snowy egret, from Audubon's *The Birds of America*, 1835.

become one of the charter members of the American Bison Society. Grinnell grew up on Audubon's former estate along the Hudson River, and spent his childhood exploring not only its woods and streams but the collections of antlers, bird specimens, and magnificent works of art amassed during Audubon's many adventures. As Grinnell grew older, he became convinced that neither professional ornithologists nor existing laws would save the birds his hero had immortalized on paper. In 1886, as the editor of the magazine *Forest and Stream*, he proposed the creation of "an Association for the protection of wild birds and their eggs, which shall be called the Audubon Society."

By the time it was incorporated the following year, the Audubon Society had attracted thirty-nine thousand members, and Grinnell, who was still editing *Forest and Stream*, launched a second magazine dedicated to the new society and its mission. But the effort was understaffed and underfunded, and Grinnell was overwhelmed by the workload and discouraged by public indifference to the excesses of the plume trade. He shuttered both the society and its magazine in 1888, writing a bitter farewell: "Fashion decrees feathers; and feathers it is."

Eight years later, in Harriet Hemenway's well-appointed Back Bay living room, her hand-picked circle of Bostonians founded the Massachusetts Audubon Society. Their fortunes and influence sustained the Audubon movement through its second infancy. By 1898, fifteen more states and the District of Columbia had established Audubon societies, and in 1905 the National Association of Audubon Societies was incorporated in New York state.

Hemenway and her Audubon allies successfully pushed for state laws restricting the feather trade, and they championed the federal Lacey Act, passed in 1900, that banned the interstate sale and transport of animals killed in violation of state laws. They faced plenty of opposition, not only from the millinery industry but from its patrons. In New Jersey in 1910, according to one press account, women opposed to a proposed plumage ban packed the ladies' gallery of the statehouse "and made themselves conspicuous by calling the assemblymen endearing names and pelting them with confetti and wads of paper."

Audubon activists celebrated in 1918 when Congress effectively ended the plume trade in the United States by passing the Migratory Bird Treaty Act. Three years later, after decades of lobbying by the Royal Society for the Protection of Birds, the British Parliament outlawed the importation of most plumage.

Over the following years, Audubon volunteers watched as bird populations recovered from the plume trade. In Florida in the early and mid-1920s, participants in the national Christmas Bird Count—an Audubon tradition inaugurated by Frank Chapman in 1900—reported total numbers of great egrets in the single digits. During the Christmas Bird Count of 1937, after nearly twenty years of the Migratory Bird Treaty Act, one intrepid birdwatcher in southwestern Florida counted more than a hundred great egrets in a single day.

The women who opposed the plume trade had much in common with Rosalie Edge. Though privileged by wealth and class, they were frustrated by the limited opportunities for women in general and stifled by the narrow expectations for women of their position. Rarely encouraged to pursue advanced educations, or professions of any kind, they were expected to be financially dependent on their parents and, later, their husbands. For many, the burgeoning women's suffrage movement opened a path out of the domestic sphere and into public life.

Edge, who was born into a prominent Manhattan family that claimed Charles Dickens as an ancestor, grew up surrounded by valued customers of the plume trade; as a child, she was given a silk bonnet wreathed with exquisitely preserved ruby-throated hummingbirds. Edge was an unusually accomplished student and, later, an unusually intrepid traveler, accompanying her husband, Charlie, when his firm assigned him to pursue railroad construction contracts in Asia. But she showed little interest in politics until, after three years abroad, she and Charlie boarded the ocean liner *Mauretania* and sailed for New York in early 1913.

Edge, then thirty-five and in the third trimester of pregnancy, spent much of the trip resting in her cabin, but during her appearances on deck she was befriended by British noblewoman Sybil Haig Thomas, the Viscountess Rhondda. Lady Rhondda was the mother of Margaret Mackworth, then one of the most visible and militant members of Britain's suffrage movement. Edge listened as Lady Rhondda, who ardently supported her daughter's cause if not her tactics, described how Mackworth and her fellow suffragists barged into men's clubs, heckled politicians, and rioted in the streets. Edge would later describe her shipboard conversations with Lady Rhondda as "the first awakening of my mind," and it would not be long before she acted on that awakening. The Edges' son, Peter, was born soon after their arrival in New York, and their daughter, Margaret, in the spring of 1915. Within a few weeks of Margaret's birth, Edge dedicated herself to the suffrage cause, and she rose quickly through the ranks of the New York State Woman Suffrage Party. Around the same time, on a whim, she wrote a check for a life membership in the National Association of Audubon Societies.

The Audubon movement, meanwhile, was at odds with itself. While its membership was unified in its opposition to plume hunting, it was divided over protections for eagles, hawks, and other birds of prey. Some Audubon members were sport hunters who saw both mammalian and avian predators as threats to their own cherished pastime, and believed that both should be controlled by humans—either through hunting or through government trapping and poisoning programs like those run by the new Bureau of Biological Survey. Many Audubon members— mostly birdwatchers, but some hunters, too—were just as passionate about birds of prey as they were about other bird species.

The members of this second faction were outraged when, in 1917, the leaders of the Audubon association declined to publicly oppose a bounty on bald eagles that had been approved by the Alaska Territorial Legislature, opting for a less confrontational approach. Eight years later, when association president Gilbert Pearson refused to openly criticize a hawk eradication campaign mounted by a national sportsmen's organization, the dissenters began to organize a rebellion.

While the Audubon movement was quarreling over eagles, New York became the first state in the eastern United States to guarantee women the right to vote. The victory, achieved in late 1917, opened the door to nationwide women's suffrage, which would become a reality when three-quarters of the states ratified the Nineteenth Amendment to the Constitution in 1920.

With suffrage in New York assured, Edge's involvement with the movement ebbed, and she turned her attention to taming Parsonage Point, a weedy, rocky, four-acre property on Long Island Sound that Charlie had purchased in 1915. In 1919, with house construction delayed by wartime shortages, the family lived on the property in tents. Every morning, Edge crept out to watch a family of kingfishers, and soon became acquainted with the local quail, kestrels, bluebirds, and herons, recording her sightings each day. While Peter and Margaret, six and four, planted pansies in the garden, Edge decorated the trees and shrubs with suet and scattered birdseed on the ground.

In spite of their joint endeavors at Parsonage Point, Edge and her husband drifted apart. After an argument one evening in the spring of 1921, Rosalie left with the two children for her brownstone on the Upper East Side. The Edges did not divorce, but they eventually secured a legal separation, which both avoided the scandal of a public divorce and required Charlie to support Rosalie with a monthly allowance—which he reliably did. For Rosalie, however, the split was devastating. She later described it as "my earthquake," and mourned not only the loss of her husband but the loss of her home at Parsonage Point—"the air, the sky, the gulls flying high."

For more than a year, Edge took little notice of the birds around her, in Manhattan or anywhere else. But in late 1922, she began to make sporadic notes on the species she saw in the city. Three years later, on a warm May evening, she was sitting by an open window in her brownstone when she heard the sharp staccato shriek of a nighthawk and looked up. Years later, she would muse that birdwatching "comes perhaps as a solace in sorrow and loneliness, or gives peace to some soul wracked with pain."

Edge began birding in nearby Central Park, often with her children and red chow in tow. She soon learned that the park was at least as rich in bird life as Parsonage Point, with some two hundred species recorded there every year; a wooded island in an urban sea, the park was, and is, a critical rest stop for migratory species and a valuable refuge for resident birds. At first, Edge's noisy entourage and naïve enthusiasm irritated the park's rather shy and clannish community of bird enthusiasts. She was a quick learner, however, and she found a willing mentor; she began checking the notes that Ludlow Griscom, then the American Museum of Natural History's curator of birds, left for other birders in a hollow tree each morning, and soon she befriended the man himself. Young Peter shared her renewed passion for birdwatching, and, as she grew more knowledgeable, she would call his school during the day with instructions about what to look for during his walk home. (When the school refused to pass on any more phone messages, she started sending telegrams.) She gained the respect of park birders, and in the summer of 1929, one of them mailed her a copy of the sixteen-page pamphlet called "A Crisis in Conservation."

Edge received the pamphlet at the Paris hotel where she was ending a summer tour of Europe with her children. Though she didn't recognize the return address on the envelope, the dire warnings inside caught her attention. "Let us face facts now rather than annihilation of many of our native birds later," she read. "Our success is far from complete; in many cases there has been no success at all and no sincere effort is being made to achieve any."

The authors argued that large bird protection organizations—Audubon was not named, but clearly implicated—had been captured by gun and ammunition makers, and were failing to safeguard the bald eagle and other species that many sportsmen considered pests or targets.

In her brocaded hotel room, Edge read and reread the pamphlet. "I paced up and down, heedless that my family was waiting to go to dinner," she recalled. "For what to me were dinner and the boulevards of Paris when my mind was filled with the tragedy of beautiful birds, disappearing through the neglect and indifference of those who had at their disposal wealth beyond avarice with which these creatures might be saved?" A few days later, after Edge and her family boarded an ocean liner for their return journey to New York, she settled into a deck chair and continued to pore over the pamphlet's claims.

When Edge and her children were back on the Upper East Side, she asked her birding friends about the pamphlet, and they suggested she contact Willard Van Name, one of the authors. Van Name, an invertebrate zoologist at the Museum of Natural History, was known for his frequent, closely typed letters of complaint about his fellow scientists and the direction of the conservation movement. Though Hornaday applauded his critiques, historian Stephen Fox writes that most conservationists dismissed him as a "quixotic, truculent curiosity." After the publication of "A Crisis in Conservation," Gilbert Pearson wrote to museum director Henry Fairfield Osborn—the New York Zoological Society co-founder and fellow-traveler of Madison Grant—complaining that the pamphlet was "worthy of a place with the Lamentations of Jeremiah." Osborn's administration quickly denounced the pamphlet and demanded that Van Name sign a new contract restricting his publications to those about "ascidians and isopods." George Bird Grinnell, who served on the board of the national Audubon association, counseled his colleagues to ignore the pamphlet, and they readily took his advice.

Edge, however, was unaware of Van Name's reputation. When they met for a walk in Central Park, she was impressed by his knowledge of birds—he could identify them by plumage and call—and his dedication to the conservation of both animal and plant species. Van Name, who had grown up in New Haven, Connecticut, in a tight-knit family of

Yale University scholars, had a particular fondness for trees, which he had learned to appreciate during childhood walks with his uncle. "No more pernicious nonsense can be disseminated," he wrote in a scathing 1925 letter to *Science*, "than the idea that if we do not hurry up and cut the rest of our dwindling supply of timber the forests are going to fall down and rot like a crop of weeds."

Van Name was a lifelong bachelor and confirmed misanthrope, preferring the company of trees and birds to that of people. As with Hornaday, however, Edge discovered that Van Name had a "sweet and gentle streak," and their introductory walk in the park was the first of many. Though Van Name rarely had a good word for members of his own species, he would come to describe Edge as "the only honest, unselfish, indomitable hellcat in the history of conservation."

Van Name convinced Edge that his pamphlet's arguments were sound, strengthening her resolve to confront the Audubon association directors. When Edge visited Hornaday for the first time, he further confirmed the claims made in "A Crisis in Conservation." Hornaday had no great love for eagles and other birds of prey—there were times, he argued, when it was "fair and right to go after them with guns and reduce their numbers"—but, like Van Name, he was enraged by what he saw as the corruption of the Audubon association by commercial interests, and the resulting quietism in bird conservation as a whole.

Hornaday told Edge that years earlier, association president Pearson had entertained a large donation from gun and ammunition makers— a gift that would have included a doubling of Pearson's salary—and that the association declined the donation only after withering public criticism from Hornaday and others. (The Audubon association retreated with some bitterness, implying that Hornaday, who was raising money for the New York Zoological Society at the time, was simply envious.) But the association continued to collaborate with gun makers, and Hornaday, whose disdain for commercial and subsistence hunting had expanded to include most sport hunting, believed that Pearson and the organization's agenda were heavily influenced by such supporters.

Edge, fresh from the vibrant, often volatile women's suffrage movement, was shocked by this description of acquiescence.

A few days after meeting with Hornaday, Edge wrote a courteous letter to Pearson, reiterating her questions about the pamphlet and asking
once again if the association planned to address its charges. Pearson
responded, in part, by saying that birds of prey such as the bald eagle
were so reviled that there was no hope of protecting them. Audubon
would not be trying to persuade the government to restrict hunting
in national wildlife refuges, he said. No one conservation organization
could do everything. And no, the Audubon association would not be
answering the accusations in Van Name's pamphlet.

That fall, as the country entered the emergency of the Great Depression, Van Name and Edge spent many evenings together in the library
of her brownstone on the Upper East Side, plotting their response to
what they saw as a correspondingly dire emergency in conservation.
Between puffs on his long cigars, Van Name offered suggestions as
Edge typed up their latest fierce letter. The prickly scientist became
such a fixture in the household that he began to help Margaret with her
algebra homework, and taught Peter how to carve a roast duck with sur-
gical precision. Edge, with typical bravado, named their new partner-
ship the Emergency Conservation Committee—a title that exaggerated
the group's numbers but foreshadowed its outsized impact.

In December of 1929, about six hundred bird enthusiasts received a new
pamphlet, elaborately titled "Framing the Birds of Prey: An Arraignment
of the Fanatical and Economically Harmful Campaign of Extermination
Being Waged Against the Hawks and Owls." Neither Van Name nor
Edge signed it, but it was their first joint publication. Over the next twelve
months, they co-wrote five more pamphlets, including another direct shot
at the Audubon association and its neglect of birds of prey. Articulate and
colorfully written, the pamphlets placed blame and named names, and
they were immediately popular. Requests for additional copies poured in,
and Edge and Van Name mailed them out by the hundreds.

Their challenge to the still insular conservation establishment did not go unanswered. Pearson kept a file of supportive letters from Boone and Crockett types, and Mabel Osgood Wright, an early leader of the Audubon movement, responded to one Emergency Conservation Committee attack by calling Van Name a "sorehead" whose "too narrow life and much brooding" had led to a "serious mental warp."

During the committee's first year, Edge recruited a young newspaperman named Irving Brant, who had become disillusioned with the Audubon association while writing about its opposition to tighter hunting limits on national wildlife refuges. For the next three decades, Edge, Van Name, and Brant would serve as the core—and often the only—members of the committee. Usually, they preferred it that way. "[We] could strike hard on any issue without being toned down," Brant remembered. Edge and Brant contributed their skill with words, while Van Name, who paid for the printing, contributed his scientific expertise.

The woman once dismissed by an Audubon attorney as a "common scold" eventually won her war with the association. When the Audubon leaders denied Edge access to the association's list of members, she took them to court and prevailed. In 1934, faced with a declining and restive membership, Pearson resigned, and over time the association, renamed the National Audubon Society, distanced itself from supporters of predator control and instead embraced protection for all bird species, including birds of prey. "The National Audubon Society recovered its virginity," Brant wryly recalled in his memoir. Today, while the nearly five hundred local Audubon chapters coordinate with and receive financial support from the National Audubon Society, the chapters are legally independent organizations, and they retain a grassroots feistiness recalling that of Rosalie Edge.

The Emergency Conservation Committee would last for thirty-two years, through the Great Depression, the Second World War, five presidential administrations, and frequent quarrels between Edge and Van Name. It published dozens of pamphlets, and was instrumental in not only reforming the Audubon movement but establishing Olympic and Kings Canyon national parks and increasing public support for conservation in general. Brant, who later became a confidant of Harold Ickes, Franklin Roosevelt's Secretary of the Interior, remembered that Ickes would occasionally say of a new initiative, "Won't you ask Mrs. Edge to put out something on this?"

In 1933, only a few years after her confrontation with Pearson at the Audubon meeting, Edge encountered a longstanding Pennsylvania tradition: each fall, recreational hunters shot thousands of migrating hawks, both for sport and to reduce what was believed to be rampant hawk predation on livestock. Hunters had learned that above one exposed ridgetop in southeastern Pennsylvania, topography and wind currents narrowed the hawks' migration path, concentrating the birds so that it was possible to kill thousands in the course of a few weekends. When Edge saw a photograph of more than two hundred hawk carcasses lined up on the forest floor, she was horrified, and when she learned that the ridgetop and its surrounding land was for sale, she became determined to buy it.

In the summer of 1934, she signed a two-year lease on the property, reserving an option to purchase it for about three thousand dollars—once again clashing with the Audubon association, which also wanted to buy the land. "Practically everything in the world is rentable, from a dress suit to an ocean liner," she said. "Why not a mountain?" Van Name loaned her the five hundred dollars she needed for the lease, and she raised funds for the purchase from supporters.

Edge, contemplating her new real estate, knew that fences and signs would not be enough to stop the seasonal hawk hunt; she would have to hire a warden. "It is a job that needs some courage," she cautioned when she offered the position to a young Boston naturalist named Maurice Broun. Wardens charged with keeping plume hunters out of Audubon refuges faced frequent threats and harassment, and one of them—Guy Bradley, who guarded rookeries in southern Florida—had been murdered by poachers in 1905. Broun, though newly married, was not dissuaded, and he agreed to start work that fall. Edge, in classic fashion, assigned him temporary authority over the property and promptly left on a steamer for Panama.

When Broun and his wife, Irma, arrived in Pennsylvania, they had to ask directions to the mountain they were supposed to guard. The locals, many of whom spoke the German dialect known as Pennsylvania Dutch, greeted the Brouns with suspicion but grudgingly respected the property rights the couple represented, and gradually the shooting stopped—though Irma patrolled the road to the sanctuary for many years, greeting and approving every visitor. At Edge's suggestion, Broun began to make daily counts of the birds that passed over the mountain each fall, beginning a tradition that was interrupted only by his deployment to the Pacific during the Second World War.

Hawk Mountain has since accumulated the longest and most complete record of raptor migration in the world. From these data, research-

Irma Broun, local conservationist Clayton Hoff, Rosalie Edge, and Maurice Broun at the entrance to Hawk Mountain Sanctuary, circa 1940.

ers know that golden eagles are more numerous along the flyway than they used to be, that sharp-shinned hawks and red-tailed hawks are less frequent passersby, and that kestrels, the smallest falcons in North America, are in steep decline. And Hawk Mountain is no longer the only window on raptor migration; there are some two hundred active raptor count sites in North and South America, Europe, and Asia, many founded by the international students who train at Hawk Mountain every year. Biologists study these lengthening datasets for early signs of trouble within populations and species, mindful of the advice of Willard Van Name: "The time to protect a species is while it is still common."

Broun usually counted hawks from North Lookout, a pile of sharp-edged granite on the rounded peak of Hawk Mountain, and one frosty October morning in 2018 I climbed the winding half-mile path that leads to the lookout. Laurie Goodrich, the director of conservation science at Hawk Mountain Sanctuary, was already on watch, staring down the ridge as a chilly wind swept in from the northwest. She has been scanning this horizon since 1984, and during some years has put in more than a hundred twelve-hour days at the lookout. The view is as familiar to her as a spouse's face.

"Bird coming in, naked eye, slope of Five," Goodrich said to her assistant, using the landmark nicknames established by Broun. A sharp-shinned hawk popped up from the valley below, racing by just above our heads. Another followed, then two more. A Cooper's hawk zoomed toward us, taking a swipe at the horned-owl decoy perched on a wooden pole nearby. Goodrich, whose bright eyes and sharp features give her a passing resemblance to the birds she watches, seemed to be looking everywhere at once, calmly calling out numbers and species names as she greeted arriving visitors.

Like the hawks, the birdwatchers arrived either alone or in pairs. Each found a spot in the rocks, placed Thermoses and binoculars within easy reach, and settled in for the show, bundling up against the wind. By ten o'clock, more than two dozen birders were at the lookout, arrayed on the rocks like sports fans on bleachers.

Suddenly, all of them gasped; a peregrine falcon, belly full of breakfast, was barreling along the ridge toward the crowd. Peregrines, which were almost extinguished in the United States by widespread use of the pesticide DDT, have recovered since the 1970s, thanks to both pesticide regulation and the release of captive-bred birds. But they are still a relatively rare sight at the lookout, and the crowd was thrilled.

Though the ridge was quiet, with only the murmurs of conversation and a distant train whistle competing with the wind, all around us was motion and drama. Our perch was so high and exposed that many birds flew below us, and the effect was intoxicating. Earlier in the week, some of these birds had surely flown over Manhattan, touching down in Central Park to delight the park's still devoted corps of birders. Some would continue south as far as Peru.

By the end of her day at the lookout, Goodrich had greeted several dozen birders and a flock of sixty chatty middle-schoolers. She and her two assistants—one from Switzerland, the other from the Republic of Georgia—had also counted two red-shouldered hawks, four harriers, five peregrine falcons, eight kestrels, eight black vultures, ten merlins, thirteen turkey vultures, thirty-four red-tailed hawks, thirty-four Cooper's hawks, thirty-nine bald eagles, and one hundred and eighty-six sharp-shinned hawks. It was a good day, but then again, she said, most days are.

When Goodrich first arrived at the sanctuary, Jim and Dottie Brett helped unload her belongings from her pickup truck. Jim Brett, who served as a sanctuary curator for twenty-five years, grew up nearby, and as a boy helped Broun with chores around the property. He also met Rosalie Edge. "She was just a miraculous woman," he told me. "She put a foothold in this mountain and didn't let go. She also didn't put up with any crap—including from me, her young employee."

After Jim retired, the Bretts moved from a house at the sanctuary to a two-story log cabin near the foot of Hawk Mountain. Jim fondly remembers his thousands of conversations on North Lookout, many with eminent scientists and conservationists. "There were always some new luminaries there to spruce things up in your mind," he told me. "Birds almost became secondary to the whole odyssey—it was the people you were surrounded by."

Ernst Mayr, the German-born ornithologist whose definition of a species as a group of interbreeding organisms remains the most commonly accepted today, visited Hawk Mountain many times and was a dedicated supporter of the sanctuary. In October 1940, even Audubon president emeritus Pearson paid a visit, passing time with the Brouns and noting the "youthful enthusiasm" of a group of visiting students. "I was impressed with the great usefulness of your undertaking," he wrote to Edge. "You certainly are to be commended for carrying through to success this laudable dream of yours." He enclosed a check for two dollars—the sanctuary membership fee at the time—and asked to be enrolled as a member.

In late 1962, less than three weeks before her death, Edge attended one last Audubon gathering, showing up more or less unannounced at the National Audubon Society annual meeting in Corpus Christi, Texas. Edge was eighty-five and physically frail, but still capable of terrorizing the Audubon leadership if she chose. With some trepidation, society president Carl Bucheister invited her to sit on the dais with him and other dignitaries during the society's banquet. When Bucheister led her to her seat and announced her name, the audience—twelve hundred bird lovers strong—gave her a standing ovation.

When Edge and Van Name founded the Emergency Conservation Committee, the language of ecology was still unfamiliar, even within the conservation movement. The concept of the food chain, sometimes called the food web, had been proposed only three years earlier by the British ecologist Charles Elton. The word "ecosystem," commonly used in ecology and conservation to describe an assemblage of interacting species and their physical surroundings, would not be coined until

1935. Many scientists—and most of the general public—continued to think of the living world as an assembly of relatively independent parts, not an interconnected whole.

Edge's understanding of ecological relationships, which she owed to Van Name's tutelage and her own sharp intelligence, set her apart from most conservationists of her time. Her concern for all species and her opposition to most hunting were shared by animal welfare activists, including many of the women who opposed the plume trade. But while Edge hated cruelty to individual animals, she devoted most of her energy to preventing the extinction of species—placing her in the company of conservation-minded ornithologists like Alfred Newton, who welcomed support from "sentimentalists" but worried that their emphasis on individual suffering would blind the public to the gravity of extinction. Edge was also somewhat more sympathetic to sportsmen than her rhetoric suggested, for she believed that limited hunting and trapping of certain species could sometimes benefit an ecosystem. She had at least a sliver of common ground with all, but she allied herself with very few—preferring, like Hornaday, to follow her distinctive personal agenda.

In May 1934, nearly five years into the Great Depression and a month before Edge leased the land that would become the Hawk Mountain Sanctuary, the sky over Manhattan darkened hours before sunset, and a gritty wind howled through Central Park. The topsoil of Oklahoma, loosened by drought and overuse, was blowing across the continent and into Edge's dining room. "I wish now that I had kept the contents of my vacuum cleaner," she mused later, for they "would have been interesting to analyze." Edge, alarmed by the storm's implications, called Hornaday, who was beginning the physical decline that would end with his death three years later. "Thank God," he said. "Now the people have seen it, they will understand."

Edge knew what he meant, for her ecological education had confirmed what she instinctively believed to be the whole point of species conservation. "The birds and animals must be protected," she wrote, "not merely because this species or another is interesting to some group of biologists, but because each is a link in a living chain that leads back to the mother of every living thing on land, the living soil."

The following month in Madison, Wisconsin, the state university invited its first and so far only professor of wildlife management, Aldo Leopold, to help dedicate its new arboretum. Leopold, too, was deeply disturbed by the dust storms boiling up from the plains, particularly because the phenomenon called the Dust Bowl was a disaster he had foreseen years earlier, as a young forester in the U.S. Southwest. Though no exact transcript of Leopold's dedication speech survives, he turned his remarks into an essay published a few months later. The "philosophical imperialism" of humans toward plants, animals, and soil, he wrote, had led to "ecological destruction on a scale almost geological in magnitude." Attempts to stop the sabotage, he added with uncharacteristic bite, had so far accomplished little: "There is a feeble minority called conservationists, who are indignant about something. They are just beginning to realize that their task involves the reorganization of society, rather than the passage of some fish and game laws."

Leopold would devote his remaining years—and his considerable gifts—to envisioning this reorganization, using the language of ecology to describe a new relationship between humans and the rest of life. His efforts would, in their way, unsettle the institutions he worked within as thoroughly as the Emergency Conservation Committee had unsettled the Audubon association.

Chapter Four

THE FORESTER AND
THE GREEN FIRE

The slump-shouldered wooden shack on the floodplain of the Wisconsin River doesn't look like much, and in most respects, it's not. The former chicken coop, its unpainted siding patched and repatched, stands in the generous shade of an old sugar maple, its front door flapping open and a small, square window gazing at arriving visitors like a gently watchful eye. Behind the shack, at a demure distance, lists a narrow wooden outhouse.

I first visited this shack on a muggy summer morning, in the company of about a dozen conservation biologists from all over North America. We filed through the door and crammed inside, where we inspected the stone fireplace, the pine roofbeams, and the set of wooden bunks. Chipped enamel bowls hung on the wall, along with a well-worn cross-cut saw. Two kerosene lamps were tucked into a corner, below an old-fashioned wire toaster made for fireplace use. "I was raised Catholic, so I get it," one scientist said with amusement. "We have to see the relics."

The bowls, the lamps, the toaster: they were relics, and yet not. There were no velvet ropes inside the shack, no silent guards. Visitors lounge on the benches, perch on the hearth, and sometimes break the glasses. A half-used roll of paper towels hung just inside the door. "We try to keep it to Leopold standards," our guide said. "Which are not very high."

In 1935, a few months after his impassioned speech at the University of Wisconsin Arboretum, Aldo Leopold decided to invest in a piece of land. He wanted a place where he could hunt, his children could play, his wife, Estella, could hone her championship archery skills, and the whole family could undertake a conservation experiment of its own.

He poked around the countryside near Madison and found eighty abandoned acres along the Wisconsin River, priced for a professor at eight dollars an acre. The land had been home to the Ho-Chunk people until most were forced out by a lopsided 1837 treaty, then to white settlers until most were forced out by the Dust Bowl. The sandy soil was exhausted, and the chicken coop was a wreck piled with manure. The mosquitoes were ferocious. But Leopold was jubilant about the land's "opportunities," and soon the rest of the family loved it as much as he did.

During weekend visits, the Leopolds cleaned the inside of the coop and installed a clay floor. They built a bunkroom out of salvaged lumber, nicknamed the West Wing, and the outhouse, which they called the Parthenon. (The youngest Leopold, also named Estella and known as Estella Jr., was shocked to learn in school that there was *another* Parthenon.) Despite the improvements, the Leopold children once heard a cry of amazement from a passing car: "Look, Mother, someone *lives* there!"

Over the next decade and a half, the family planted trees—some three thousand pine seedlings every year. "All right, so now you *grow!*" Estella Sr. would say, with mock sternness, to each newly settled pine. At first, most of the pines died, but after the drought eased in 1940, the young trees took hold. Leopold also started planting a scrap of prairie, using the crude but effective method of digging up a square yard of native grasses elsewhere, plopping the chunk of dirt on the roof of his car, and relocating it on his land. The university arboretum's prairie restoration project is the oldest in the world. The Leopolds' is the second oldest.

While Leopold shoveled manure, planted the eroded riverbanks, and

"Look, Mother, someone *lives* there!": Aldo, Estella, Estella Jr., and Starker at the Leopold shack, 1938.

turned worn-out cornfields back to prairie, he took note of the species growing, singing, and moving around him, and he encouraged his children to do the same. In the very early mornings, while sitting by the campfire with a coffeepot on the boil, he listened as the songbirds started their day, using a meter to measure the sunlight each time a species joined the chorus. His observations accumulated in a series of shack journals—becoming the raw material for some of conservation's most famous texts.

Aldo Leopold is the kind of author whom people cite as an influence even if they've never picked up one of his books. He was almost dangerously eloquent; among the usual pantheon of North American conservation writers, perhaps only Henry David Thoreau and Sierra Club founder John Muir are as frequently quoted and misquoted. "To keep every cog and wheel is the first precaution of intelligent tinkering";

"One of the penalties of an ecological education is that one lives alone in a world of wounds"; "I am glad I shall never be young without wild country to be young in"—these Leopoldisms, and others, have been proverbial for at least three generations of conservationists. Leopold's best-known work, the collection of essays published posthumously in 1949 as *A Sand County Almanac*, has sold more than two million copies and been translated into fourteen languages, including Latvian, Korean, and Turkish.

Though Leopold's language was often simple, his thinking was not. "His conviction was that conservation had to rest on a base that included not only the natural sciences, but also philosophy, ethics, history, and literature," writes Leopold biographer and scholar Curt Meine, and Leopold's reading and writing reflected that conviction. His ideas changed over time, and it's not unusual to hear conservationists on opposing sides of internal debates lobbing Leopold quotes at one another, each certain that the great man is an ally.

But over the course of his life, through thousands upon thousands of published and unpublished pages, Leopold burnished two bedrock principles of species conservation. The first is that each species requires a combination of space, food, and shelter that ecologists call habitat, and which Leopold referred to as land. The second is that all species need predators—even, in most cases, the predators themselves. And neither habitat nor predators, Leopold maintained, could be adequately protected by laws alone.

When Leopold was born, in January of 1887 in Burlington, Iowa, William Hornaday was building his bison display at the National Museum in Washington, DC, and George Bird Grinnell was trying to keep up with the success of his new Audubon Society. On the plains, the slaughter of the bison was still underway, and would soon lead to new human tragedies.

After the dry, hot summer of 1890, the Lakota Sioux—many of whom had switched to agriculture as bison numbers dropped—faced both a failed harvest and radical reductions in government rations. Desperate, some began to perform the Ghost Dance, a ritual that the Paiute spiritual leader Wovoka promised would bring a new world into being. Government agents misinterpreted the dance as a prelude to battle, leading to a standoff that ended with the massacre of more than one hundred and fifty Lakota Sioux by U.S. soldiers on the banks of Wounded Knee Creek.

The same year, the superintendent of the national census declared—on somewhat shaky evidence—that the western boundary of white settlement had progressed all the way across the continent to the Pacific Ocean. The so-called "frontier line" was gone. This announcement excited far more concern for the conquerors than the conquered, especially after historian Frederick Jackson Turner proposed that the frontier—which he called "the meeting point between savagery and civilization"—had inclined the "American character" toward democracy and individualism. Turner's thesis, adopted and popularized by Theodore Roosevelt and others, fed urban elites' fears about the future of white male vitality: if the frontier had forged the national character, what would become of the national character without it?

On a high limestone bluff on the western bank of the Mississippi River, in a house built by his grandfather Charles Starker, Aldo Leopold and his three younger siblings grew up shielded from the savageries of civilization. Much like young Hornaday thirty years earlier, Leopold roamed the Iowa countryside, often searching for birds. When his parents gave him a copy of Frank Chapman's *Handbook of Birds of Eastern North America,* it cemented his habit of describing the world around him. "From my own observations I have found that during the nesting season when the young are dependent, about six weeks, the old wrens carry insects into the box for twelve hours each day at the rate of five times every ten minutes," fourteen-year-old Leopold wrote in his school composition book. He added, "Nothing could be more suggestive of complete content and happiness than the song of the wren."

Leopold's first conservation mentor was his father, a former traveling salesman who had peddled products ranging from roller skates to barbed wire before settling down to run a desk manufacturing company in Burlington. Carl Leopold, who came from a large family of German immigrants, had no formal training in conservation or science, but he had hunted since he was a boy, and over the years he had watched duck numbers drop. Though local restrictions on which species could be hunted, and when, dated back to colonial times, states had only begun to limit how many animals could be killed at a time—Iowa was the first state to do so, in 1878—and those rules tended to be generous to a fault. The elder Leopold understood that hunters needed to restrain themselves in order to preserve their sport, so he strictly limited the number and kinds of birds he killed. And while he delighted in a successful hunt, he sometimes left his gun at home, making it clear to his young sons that time outdoors was more important than the chase. For Aldo, never a regular churchgoer, these adventures were his Sunday school.

In 1904, with the encouragement of his mother, Clara, Aldo left Burlington for preparatory school in New Jersey. His chatty letters home detailed his almost daily tramps beyond campus: "I went north, across the country, about seven miles, and then circled back toward the west," he wrote to his mother soon after his arrival in January. He described sightings of more than a dozen bird species, concluding that he was "more than pleased with the country."

His classmates thought Leopold eccentric, but they admired his enthusiasm, and they sometimes joined his outings. He was shy, with an internal reserve that was sometimes taken for haughtiness, but his new friends loosened him up. Photographs from the time show a young man with prematurely receding blond hair and a face that was forbidding when serious but could be transformed by his warm, wide grin.

Though Leopold steadily checked new bird species off his list—"I am now acquainted with 274 species of birds in the United States"— his interests went beyond Linnaean classification, which he would later dismiss as "baptizing species and describing feathers and bones." He noted that goldfinches and pine siskins fed together, and that phoebes tended to gather around blooming heads of skunk cabbage. He was awakened to the wonders of plants by Asa Gray's *Manual of Botany* (later a well-thumbed reference at the Leopold shack) and earnestly pronounced Charles Darwin's *The Formation of Vegetable Mould Through the Action of Worms* to be "of much interest and surprise." He was curious about other species, but he was fascinated by the relationships among them.

For a young man like Leopold—gripped since childhood by what he jokingly called "woods fever," inclined toward science but idealistic enough to aspire to national service—the obvious place to further his education was Yale University, where pioneering forester Gifford Pinchot had co-founded the country's first graduate school of forestry in 1900. When Leopold arrived at Yale in the fall of 1905, Pinchot was serving as chief of the U.S. Forest Service under President Theodore Roosevelt, overseeing a massive expansion of the national forest system.

Pinchot is usually remembered for his very public quarrel with John Muir, his onetime friend and mentor, over San Francisco's plans to build a dam and reservoir in the Hetch Hetchy Valley in Yosemite National Park. Muir, the survivor of a rigidly disciplined childhood in Scotland and Wisconsin, awoke to the wonders of the world around him after being temporarily blinded in his twenties, and for the rest of his life he was happiest in California's Sierra Nevada, traveling by foot and carrying little more than a walking stick.

In the spring of 1903, Roosevelt—who, like many others, was moved by Muir's poetic descriptions of the Sierra—requested his company on a camping trip in Yosemite. "I do not want anyone with me but you," he wrote. Roosevelt and Muir got off to a shaky start, for neither wanted to let the other get a word in edgewise; Muir made things worse by trying to decorate the president's lapel with twigs. But they warmed to each other, and their conversations around the campfire helped persuade Roosevelt to strengthen protections for Yosemite and to create the dozens of national parks, wildlife refuges, and monuments that became part of his legacy. They may also have led Roosevelt to question his devotion to hunting. "Mr. Roosevelt, when are you going to get beyond the boyishness of killing things?" Muir recalled asking one evening. "Are you not getting far enough along to leave that off?" Roosevelt replied, "Muir, I guess you are right."

While Muir lost the battle over the Hetch Hetchy dam—it was approved in 1913, and stands today—he won history's sympathy, and his paeans to the California mountains became part of conservation scripture. "I used to envy the father of our race, dwelling as he did in contact with the new-made fields and plants of Eden; but I do so no more, because I have discovered that I also live in 'creation's dawn,'" Muir reflected during one trip. "The morning stars still sing together, and the world, not yet half made, becomes more beautiful every day."

Pinchot, who prized the efficient use of natural resources, is often caricatured as a bureaucrat preoccupied with board-feet. His conflict with Muir is considered the archetypal cleavage between utilitarians and preservationists—between those who primarily want to maintain landscapes and species for people, and those who want to protect them from most human use. Early in his career, Pinchot did tend to treat forests like commodities, much as Hornaday treated bison as livestock, but his utilitarianism was rooted in concern for the future, with the goal of "the greatest good, for the greatest number, for the longest run." His views evolved; by 1920, he was speaking of forests as a "living society of living beings," and decrying the cozy relationship between the Forest Service and the timber industry. In the 1930s, he became a committed internationalist, arguing that conservation and global peace were as interdependent as humans and the rest of life.

Pinchot was less quotable than Muir, but he was ultimately broader-minded, concerned with the well-being of all species. When Muir, on one of his hikes in the Sierra, encountered a group of Mono people—whose ancestors had lived in the area for more than a thousand years—he mused that they "seemed to have no right place in the landscape," and struggled with his aversion to their unwashed faces. His sense of the mountains as a purifying refuge from civilization was embraced by Madison Grant, whose circle of wealthy sportsmen sought to preserve the California redwoods as an eternal frontier for the fair-skinned elite. Pinchot, meanwhile, often argued that forests and other landscapes should be managed "for the benefit of all the people instead of merely for the profit of a few."

Right: Theodore Roosevelt and John Muir on Glacier Point above the Yosemite Valley, May 1903.

Left: Theodore Roosevelt and Gifford Pinchot on the river steamer *Mississippi*, October 1907.

Along with Grant, Roosevelt, and many other affluent intellectuals of his time, Pinchot was a supporter of eugenics, the practice of "improving" humanity through various controls on reproduction. But unlike Grant, who saw eugenics as a means of "race betterment," Pinchot believed it could ease poverty—which he viewed as a kind of pollution that harmed all life. Pinchot's most valuable legacy, contends his biographer Char Miller, lies not in his role as Muir's foil but in his "effort to reach an ever more complex understanding of the tangled relationship between humanity and the natural world in which it exists."

Like Leopold, Pinchot was the adored eldest son of a devoted and determined mother, but he was born into much loftier circumstances. His father, James Pinchot, had earned a fortune in the logging business and, while still in his thirties, become a patron of American landscape art (he named his son after Sanford Gifford, a member of the Hudson River School of landscape artists). When the younger Pinchot came of age, his father suggested that he pursue a career in forestry. It was a curious piece of advice, for, at the time, the profession of forestry barely existed in the United States. Logging, like hunting, was largely unregulated, and aside from a few ad hoc efforts, little thought was given to managing forests for the long term.

Gifford Pinchot, however, was not intimidated by new territory, and after graduating from Yale he went to Europe, determined to learn what he could from its forestry schools. He came home convinced that some European-style public ownership of forests was necessary to fend off short-term profiteers, but that North American foresters should avoid the fussy precision of some of their European counterparts. The public forests of the United States, he wrote to his father, should be managed so that they "will be as picturesque as though left wholly alone, and . . . bring a respectable income at the same time." His father must have appreciated this neat reconciliation of his own interests in profit and beauty.

Pinchot was not the first in the United States to propose to manage the nation's forests with the future in mind, but he was the most energetic promoter of the idea. In 1898, Pinchot was appointed head of the Division of Forestry within the U.S. Department of Agriculture, where he supervised a staff of sixty and a few thousand acres of public forests. He immediately began preparing for larger responsibilities, an optimism rewarded when his friend (and sometime wrestling opponent) Theodore Roosevelt became president in 1901. As Roosevelt expanded the national forests, Pinchot supplied the staff needed to manage them, and soon the newly christened Forest Service had five hundred employees.

This bureaucracy quickly ran into resistance in the western United States, where public lands—forested and otherwise—were valued chiefly as a source of free food for gigantic and unchecked herds of cattle and sheep. In 1907 a group of Western congressmen, determined to keep public lands out of the hands of "dreamers and theorists" in Washington, succeeded in passing strict limits on the further expansion of forest reserves.

With the bill set to be signed, Roosevelt and Pinchot made plans to designate another sixteen million acres of public land as national forest, and Forest Service field and office staff worked around the clock to map new forest boundaries. When these "midnight reserves" were announced, as Roosevelt gleefully remembered, opponents "turned handsprings in their wrath." (Wrath among Westerners over perceived federal overreach was not universal, and it was highly selective— there was little opposition to government-supported dam projects, for instance—but it would continue to simmer, and even today periodically boils into violence.)

Pinchot's efforts to create a national corps of professional foresters— "American foresters trained by Americans in American ways for the work ahead in American forests"—were greatly assisted by his parents, who in 1900 funded the founding of the Yale school. As a forestry student at Yale, Leopold absorbed the Pinchot doctrine, learning both its philosophy and its meticulous methods of survey and measurement. From the start, he showed signs of restlessness. "I am getting narrow as a clam with all this technical work," he complained to his family. Like most of his classmates, though, he aspired to a career as a forest supervisor with the emboldened Forest Service. When a fellow student declared that "I'd rather be a Supervisor than be the King of England," Leopold enthusiastically agreed.

After graduating from Yale with a master's degree in 1909, Leopold was hired to lead a six-man Forest Service reconnaissance crew in the newly designated Apache National Forest, in eastern Arizona Territory. Named for the people who had been forcibly removed from the forest by U.S. troops, the landscape was nearly roadless, and the Blue Range, where Leopold and his crew were assigned, was described by one early forest inspector as a "chaotic mass of very precipitous and rocky hills," too steep for grazing livestock. The crew traveled by foot and on horseback, surveying the range and estimating its store of standing timber.

Leopold applied himself to the job, but almost immediately ran into trouble. He got lost twice on his way to training; he ignored the advice of the more experienced men on his crew; he skimped on supplies and camp comforts, insisting that all adopt his ascetic habits. After three months of technically and physically demanding work, the crew was fed up with Leopold and one another. His supervisors, after investigating the complaints, decided to give their new hire another chance, but Leopold's Forest Service career narrowly avoided an early end.

Despite his growing pains, Leopold was entranced by his surroundings. "In the early morning a silvery veil hangs over the far away mesas and mountains—too delicate to be called a mist, too vast to be merely beautiful—it isn't describable, it has to be seen," he wrote to his mother. "I'm taking tonight off, and when I have written this I'm going out to sit on the corral fence and listen to the frogs down by the river." His second summer of reconnaissance went far more smoothly, and in 1911, at age twenty-four, he was detailed to a post in the Forest Service district office in Albuquerque.

At a spring cotillion dance in Santa Fe, he was charmed by Estella

Bergere, one of eight daughters in a locally prominent family that traced its ancestry to colonial Spain. Before Leopold had a chance to court Estella, he was again relocated by the Forest Service, this time to the Carson National Forest on the Colorado–New Mexico border. He wrote Estella rhapsodic letters about his new horse—named Polly, after a parrot-shaped lantern Estella had given him at the dance—and after a few weeks of pining, he took the train down to Santa Fe for a visit. Estella, who had other suitors, kept him wondering, and Leopold pretended not to care: "I have consistently *ignored* the absence of letters and mentioned them only when received," he wrote to his sister that summer. "And I have *not* played the doormat . . .! Nor will I. Neither shall I ever desist—but it isn't come to that yet."

After corresponding with Estella for four months and meeting her three times, Leopold proposed—and after some weeks of consideration, Estella said yes. The newlyweds moved north to the outskirts of the Carson National Forest, where Leopold had recently been promoted to forest supervisor, and Estella soon became a capable hunter and homesteader. Within a few months, the Leopolds were expecting their first child.

In the spring of 1913, during a several-day pack trip into the forest, Leopold's knees began to swell painfully, and by the time he reached home, his face, hands, arms, and neck were all noticeably inflamed. His coworkers convinced him to seek treatment in Santa Fe, and there he learned he had an acute kidney infection. He was ordered to rest for six weeks. He would not return to full-time work with the agency for another year and a half—and he would never resume the physically demanding life of a forest supervisor.

Had Leopold delayed treatment any longer, he likely would have died. Had he never fallen ill, however, we might not remember him today. During his enforced sabbatical, Aldo and Estella retreated to Iowa to await the birth of their child, a boy they would name Starker. There, away from the daily demands of the Forest Service, Leopold read widely, and began to ask himself the questions that he would later pursue in his essays and books.

Among other works, Leopold was influenced by a just-published book that he bought as a gift for his father: *Our Vanishing Wild Life*, by one William Temple Hornaday. "It is the way of Americans to feel that because game is abundant in a given place at a given time, it always will be abundant, and may therefore be slaughtered without limit," Hornaday wrote. "It is time for all men to be told in the plainest terms that there never has existed, anywhere in historic times, a volume of wild life so great that civilized man could not quickly exterminate it."

Leopold took the urgency of *Our Vanishing Wild Life* to heart. In the decades since Carl Leopold had introduced his son to his personal code of hunting conduct, the U.S. Congress had passed the Lacey Act, whose prohibition on the interstate transport of illegally acquired animals lent support to state hunting laws. President Roosevelt had established the first national wildlife refuges, beginning with a bird sanctuary on Florida's Pelican Island in 1903. But enforcement of both national and state laws was spotty, poaching was routine, and while some sportsmen practiced self-restraint, many more did not. There was also little sense, even among the country's small and scattered corps of conservation professionals, that game species—those birds and mammals hunted for sport or food—could be deliberately managed. Forestry was fast becoming a science, but game management was an afterthought, with no well-heeled Pinchots to boost its reputation and its fortunes.

When Leopold was finally well enough to return to work with the Forest Service, he put together a handbook on game management for the Southwest's forest rangers and officers—who were supposed to help state wardens arrest poachers, but rarely did. "North America, in its natural state, possessed the richest fauna in the world. Its stock of game has been reduced 98%," he wrote. "Eleven species have been already exterminated, and twenty-five more are now candidates for oblivion. Nature was a million years, or more, in developing a species . . . Man, with all his wisdom, has not evolved so much as a ground squirrel, a sparrow, or a clam." Though his facts were wobbly, his passion was unmistakable—and while Pinchot's "scientific" approach to forest management was intended to benefit wildlife as well as trees and people, such outrage on behalf of other species was not typical of the Forest Service. Leopold was begging to differ with his employer's priorities, and rearranging his own.

In the fall of 1915, Hornaday came to Albuquerque as part of a speaking tour of the western United States. He was already famous as a rescuer of the plains bison, which by then numbered in the thousands in North America, and he was beginning the skirmishes with Eastern conservation groups that would lead to his alliance with Rosalie Edge. Hornaday delivered a barnstorming lecture to an enthusiastic crowd of sportsmen, donated several hundred dollars toward the organization of a local game protection group, and signed a copy of his most recent book for Leopold: "To Mr. Aldo Leopold, on the firing line in New Mexico and Arizona." Within a few months, Leopold had founded not one but four game protection organizations in the state. As he would later tell a group of Albuquerque businessmen, the purpose of the associations was "to restore to every citizen his inalienable right to know and love the wild things of his native land."

Leopold threw himself into his new mission, despite—or maybe because of—his still shaky health. "I do not know whether I have twenty days or twenty years ahead of me," he wrote. "Whatever time I may have, I wish to accomplish something definite." In the fall of 1916, he was rewarded when the governor of New Mexico appointed the state's first chief game warden, choosing a candidate backed by the game protection associations. Hornaday reported the state's progress to "my dear Colonel" Roosevelt, and a few days later the thirty-year-old Leopold received a letter from the former president. "My dear Mr. Leopold," he wrote, "I wish to congratulate the Albuquerque Game Protective Association on what it is doing . . . It seems to me that your association in New Mexico is setting an example to the whole country."

In the spring of 1920, Leopold's interests expanded once again. He and his colleagues came home from a trip to central New Mexico, as he

wrote to his mother, "with cakes of mud a quarter of an inch thick surrounding our eyes—stuff that had blown into our eyes and was 'teared'
out so you had to pull off the lumps every few minutes."

Leopold, now second-in-command for the southwestern district
of the Forest Service, was facing the accelerating problem of soil erosion. Decades of indiscriminate grazing had stripped away much of
the landscape's vegetation, unmooring the soil, and deliberate over-
stocking to meet military demands during the First World War had
only worsened the situation. In the Gila National Forest, the riverbanks
were so severely eroded that floodwaters poured out of the mountains
more or less unimpeded, repeatedly inundating the town of Silver City
and turning its main street into a fifty-five-foot-deep ditch. Though the
Dust Bowl would not spill into Rosalie Edge's living room for another
decade and a half, Leopold saw that it was filling up.

Like Edge and her co-conspirator Willard Van Name, Leopold was
advocating for other species at a time when most North Americans and
Europeans, even those interested in conservation, thought of what we
call "the environment" as a collection of separate resources: soil, water,
animals, plants. But the science of ecology was advancing, and Leopold had a precocious understanding of ecological relationships, for he
had been observing and describing the living world since childhood.
The crumbling hillsides and riverbanks, he realized, were a shock to
both the land and its inhabitants, and any lasting remedy would require
humans to change their relationship with the land—all land, both
private and public.

Erosion also heightened Leopold's interest in protecting land as
"wilderness," a concept that was just beginning to be discussed in
conservation circles. Invoking Pinchot's doctrine that the land should
be put to its "highest use," Leopold proposed that in some cases, the
highest use was minimal use. His Forest Service colleagues were wary,
but by the spring of 1924, after several years of internal maneuvering,
he had gained enough support to complete a recreational plan for the
Gila National Forest that designated more than three-quarters of a mil-

lion acres as a wilderness area. Within the wilderness, grazing would continue at a low level, and hunting would be allowed, but no motorized vehicles would be permitted; the land would remain roadless, and humans would travel by foot or horse.

Only days after completing the plan, Leopold received word that the current Forest Service chief, William Greeley, wanted him to become the assistant director—and director-in-waiting—of the agency's Forest Products Laboratory, which Pinchot had established in Madison, Wisconsin. Though the job was a prestigious one, Leopold was far more interested in keeping trees in the ground than turning them into products. He had spent more than a decade learning the Southwestern landscape, it was Estella's beloved home territory, and the Leopolds by then had four young children. A move would not be easy. But Leopold, who had already declined several promotions and other offers that would have taken him outside the region, resolved to take the position.

On May 29, 1924, Leopold left for Wisconsin. Five days later, district forester Frank Pooler signed off on Leopold's plan for the Gila National Forest, creating the nation's first designated wilderness area. The Gila Wilderness still exists today, and in 1980, the high peaks on its eastern border were protected as the adjoining Aldo Leopold Wilderness.

Leopold spent four frustrating years at the Forest Products Laboratory, waiting (in vain) to inherit the directorship and trying (also in vain) to interest his colleagues in the possible uses of less marketable trees and forestry "waste" products like wood chips. Outside the laboratory, he spent time with his family, acquainted himself with Wisconsin conservation politics, and worked on a book about Southwestern game.

When Leopold left the Forest Service in June 1928, he turned down several job offers in favor of a contract with the Sporting Arms and Ammunition Manufacturers' Institute, an industry group that wanted him to conduct a national survey of game species. It was a less than reliable living, but it was an opportunity Leopold could not resist.

At the time, the status of game species—and indeed most other species—was a matter of guesswork. In North America, as almost everywhere else on earth, there was little precise information about how many animals there were in any particular place, how their numbers changed over time, and what they needed to survive. In the absence of data, those who wanted to protect other species were left to argue among themselves.

Hornaday, whose onetime devotion to recreational hunting had curdled into open hatred, blamed sportsmen for reported declines in the size of waterfowl flocks and deer and elk herds. Many animal welfare groups agreed with him. Many conservation groups, however, counted hunters among their founders and members; they tended to downplay the role of sport hunting in species declines, preferring to blame commercial hunting or other factors. (The Audubon association's reluctance to call for tighter restrictions on hunting had already put it on a collision course with Hornaday and Rosalie Edge.) The sporting arms institute made no secret of its vested interest in the survey's outcome, but its leaders said they wanted meaningful results, and promised to leave Leopold to his own devices.

Leopold refined his survey methods as he worked, but from the start he was interested in much more than counting heads. In each state, he interviewed hunters, foresters, farmers, reporters, and others about the game situation. All potential influences were of interest: geology, geography, vegetation, and local attitudes toward conservation and education. Though his detailed reports, written with his usual understated flair, were well received, he had barely begun to cover his territory, and at the end of the year Leopold and his sponsors agreed to limit the survey to the upper Midwest.

At the national conference of the American Game Protective Association in late 1929, only a month after the stock market crash, Leopold previewed his results. While hunting regulations were well-established throughout the upper Midwest, "control of other factors"—food, sheltering plants, predators, and disease, to name a few—was "not developed at all," and goose and duck numbers were still dropping. "The one and only thing we can do to raise a crop of game," he said, "is to make the environment more favorable . . . [This] is the fundamental truth which the conservation movement must learn if it is to attain its objective." Though fundamental, it was not nearly as obvious at the time as it might sound today. The ecological concept of habitat—the idea that each species requires a particular set of resources for survival— was relatively new. For most conservationists, protecting species still meant protecting animals from bullets, not protecting shrubbery and wetlands from bulldozers.

The conference organizers asked Leopold to help write a new policy for game protection, and his resulting report emphasized the importance of habitat. Leopold and the other members of the policy committee noted that the government had an opportunity to purchase and protect inexpensive land, as the Dust Bowl and the Depression had bankrupted many farmers. They suggested that private landowners be compensated, either by the government or individual hunters, for protecting game habitat and allowing hunting on their property.

The latter was a controversial idea. In reaction to exclusionary English game laws, many states had long allowed hunters free access to undeveloped, unfenced private land unless it was posted with signs that specifically prohibited access. But farmers were losing patience with wandering hunters, and more and more were posting their land. Paying these farmers to both conserve habitat and permit access, Leopold and his colleagues argued, would help sustain game populations and benefit both farmers and hunters. Significantly, these measures were not intended to benefit game alone; the proposed policy urged hunters to cooperate with "the non-shooting protectionist and the scientist" on "jointly formulated and jointly financed" measures protecting all wildlife.

After some heated back and forth, the policy was approved by the association, and remained in place for more than four decades. While it was the credo of a private group, not a government agency, it was the first articulation of anything like a national game conservation strategy. Leopold had argued that game was a public trust that should be managed by law, not markets; that those laws should be informed by science; and that while responsible hunters should be permitted access to game, they should also be expected to fund conservation efforts. Conservation professionals would later incorporate these ideas into what they called the North American Model of Wildlife Conservation.

The era's public concern for other species, game and otherwise, was such that Leopold's game policy brought him a degree of popular fame. He was photographed for *Time* magazine, and *American Forests* compared the adoption of the policy to a new "constitutional convention." When Leopold's final report on his game survey was published as a book, Hornaday praised it—while insisting that he would eventually be proved right about the destructive effects of all hunting. Even the impossible-to-please Willard Van Name applauded the book as a "bright spot" in his own dim view of sportsmen's organizations.

Leopold proceeded to synthesize his thinking in *Game Management*, a textbook he published in 1933. In its opening chapter, he paid tribute to an unnamed "Crusader" with a definite resemblance to Hornaday. "He insisted that our conquest of nature carried with it a moral responsibility for the perpetuation of the threatened forms of wild life," Leopold wrote, "to any one for whom wild things are something more than a pleasant diversion, it constitutes one of the milestones in moral evolution." Leopold sent a copy to Hornaday with a note of gratitude, acknowledging that "my whole venture into this field dates from your visit to Albuquerque . . . and your subsequent encouragement to stay in it."

He also asked Hornaday to help him find a job. With the game survey over, Leopold was renowned but unemployed, and the Great Depression was underway. For about a year, he and his family subsisted on his intermittent consulting work. By 1934, when Leopold gave his address at the University of Wisconsin's new arboretum, allies and admirers had convinced the university's alumni research foundation to fund his appointment as a professor of game management—the first position of its kind in the country, and perhaps the world.

Leopold took teaching seriously; for him, it was not only about mentoring young professionals but awakening students of all kinds to their relationship with the rest of life. "I shall now confide in you what the course is driving at," he told students at the midpoint of one undergraduate class. "I am trying to teach you that this alphabet of 'natural objects' (soils and rivers, birds and beasts) spells out a story." Once they learned to read the land, he told them, they would see both its beauty and its utility: "We love (and make intelligent use of) what we have learned to understand." For Leopold, there was no fundamental conflict between the romanticism of Muir and the practicality of Pinchot. He believed it was possible to love other species and use them wisely, too.

Leopold steadily introduced ecological principles into his courses and research, pushing game management beyond its agricultural foundations. Within five years, he had adopted the title Professor of Wildlife Management, and changed the name of his undergraduate game management course to Wildlife Ecology. He had also attracted an eclectic flock of talented graduate students. Frances Hamerstrom, who was a Boston debutante and worked as a fashion model before turning to science and earning her master's degree with Leopold, recalled the loyalty of his protégés: "In those days, not everybody knew that Aldo Leopold was a great man. But his students did."

The formal vocabulary of ecology was quickly becoming more sophisticated, and the reach of its practitioners was expanding. The disaster of the Dust Bowl had brought wider attention to the field, and studies of tropical forests in Panama and elsewhere in Latin America, often led by U.S. researchers, were revealing not only thousands of unfamiliar species but new levels of ecological complexity. In June 1930, New York Zoological Society scientist William Beebe—who as a young ornithologist had scouted the London feather market for Hornaday—had climbed into a cast-iron hollow globe called a bathysphere and made the first of several pioneering dives off the coast of Bermuda. Beebe eventually descended more than a half-mile below the ocean surface, becoming the first ecologist to observe deep-sea organisms in their own habitat. Ecologists were, as Leopold put it, "daily uncovering a web of interdependencies so intricate as to amaze—were he here—even Darwin himself."

Leopold's ecological inclinations were strengthened in late 1935, when he joined a group of fellow foresters on a three-month tour of Germany sponsored by an international foundation. Like Pinchot before him, Leopold was struck—and ultimately repelled—by the German forests, many of which were shaved clean of undergrowth and planted with orderly rows of spruce trees. Gamekeepers boosted deer numbers with supplemental food, which pleased wealthy hunters but created enormous herds that further damaged the forests. "One cannot travel many days in the German forests, either public or private, without being overwhelmed by the fact that artificialized game management and artificialized forestry tend to destroy each other," he wrote later. "Germany strove for maximum yields of both timber and game, and got neither."

Many Germans had also recognized the consequences of what Leopold called "cabbage brand" or "cubistic" forestry, and the country's popular *Naturschutz* movement promoted a less intrusive approach to conservation. As in the United States, however, some of the movement's leaders drew an explicit connection between the survival of other species and the survival of their own nation and race, a chilling echo of the frontier thesis that was embraced and exploited by the Third Reich. (Some prominent German conservationists—though certainly not all—in turn supported the National Conservation Law, passed only weeks before Leopold's arrival, which enabled the Reich to seize private property for nature reserves.)

Leopold and the other foresters in his group may have been only dimly aware of the political situation in Germany when they started their trip, but they surely noticed the "No Jews" signs that were by that time posted outside shops and restaurants. Leopold, whose childhood

German revived with daily practice, likely heard and saw more cruel-
ties than his colleagues did. He returned home distressed by the rise of
nationalism and militarism in Germany—and convinced that war was
inevitable. Three years later, he received an anguished letter from one
of his German hosts, who, after surviving Dachau and Buchenwald,
had escaped to Kenya, leaving a brother trapped in Germany. Leopold
appealed unsuccessfully to the U.S. State Department on the brother's
behalf, eventually arranging for him to emigrate to South Africa.

Leopold knew that the Nazi ideal of blood and soil was no answer
to the sanitized forestry he observed in Germany. What was missing
in the forests, he reflected, was "a certain exuberance which arises
from a rich variety of plants fighting with each other for a place in the
sun." The obsessive drive for order, whether in politics or the woods,
was antithetical to the emerging principles of ecology. What the land
needed was not uniformity, but complexity.

Less than a year after Leopold returned from Germany, he took the
first of two hunting trips to the Sierra Madre Occidental in northern
Mexico. There, he was plunged back into the landscape he had known
as a young Forest Service recruit, but with a difference: on the Mexican
side of the border, the forests had not yet been logged, and the riverbanks had not yet been grazed down to dust. In the mornings, flocks of
noisy parrots wheeled overhead.

The land had been inhabited by people for at least a thousand years—
by the Apache and other indigenous cultures, by mestizo communities,
by Mormon families escaping persecution in the United States—but
their numbers, and their practices, enabled them to leave a lesser mark.
It was in the Sierra Madre, Leopold later wrote, that he "first clearly realized that land is an organism, that all my life I had seen only sick land."

Leopold had traveled far in a short time, and the contrasts he witnessed led to a new essay, "A Biotic View of Land." Here he searched
for a metaphor to replace the idea of the "balance of nature," which—
though it persists in the popular imagination today—is rooted in the
outdated Linnaean concept of a centralized economy of nature, and had

Aldo Leopold bow hunting in Chihuahua, Mexico, January 1938.

already been discarded by many ecologists of Leopold's era. Borrowing from British ecologist Charles Elton, who had recently become a close friend, Leopold described the organization of living systems as a "biotic pyramid," with soil at the base and carnivores at the top. In healthy systems, he wrote, energy moves up the pyramid as organisms consume and are consumed, and down as organisms die and decay. Though these systems had always had to respond to changes in their surroundings, humans were able to effect changes "of unprecedented violence, rapidity, and scope." In this sense, efforts to protect rare species—"condors and grizzlies, prairie flora and bog flora"—were "protests against biotic violence." The ultimate purpose of conservation, in Leopold's view, was to maintain the flow of energy through living systems.

Leopold saw that this job, "the oldest task in human history," had to be undertaken at many scales. Unlike Edge and Hornaday, who at times cultivated their reputations as outsiders, Leopold was a devoted institutionalist: Yale graduate, Forest Service alumnus, university professor, Boone and Crockett Club member, scion of one large, tight-knit family and patriarch of another. He believed in the power of institutions to effect positive change, and he spent most of his life within them. But he was also a committed individualist. He rejected the growing nationalism he witnessed in Germany, and depised its U.S. strains as well. He saw conservation as the work of individuals—and each individual as "a member of a community of interdependent parts."

Leopold was serious about his work, but he was serious about his enjoyments, too. He never lost his love of outdoor adventure, and, like his father before him, he included his children in outings large and small. He tolerated, with amusement, a revolving menagerie of household pets that included several dogs, two crows, a Cooper's hawk, and a squirrel. He liked movies, highbrow and lowbrow, and he liked parties, too, though Estella sometimes had to urge him to go. He often entertained his children by walking on his hands. He was, in short, aware that life was more than thinking. "Breakfast comes before ethics," he once told his daughter Nina.

By the mid-1930s, when the Leopold family started spending weekends working and playing at the shack, Leopold's thinking about other species had completed a gradual transformation. Early in his career, he had unapologetically limited his concern to those species he considered "useful" or "interesting," and, to him, wolves and most other predators were neither. Like many other sportsmen at the time, he frequently referred to the need to reduce or eradicate "varmints" from favored hunting grounds. "It is going to take patience and money to catch the last wolf or lion in New Mexico," he told delegates to the National Game Conference in New York in 1920. "But the last one must be caught before the job can be called fully successful."

Other conservationists had already recognized that predators were part of the food web, and that their persecution tended to cause more

problems than it solved. Madison Grant had defended "so-called vermin" as early as 1911, arguing that their removal "frequently ends in
most unexpected and undesirable results." California naturalist Joseph
Grinnell, a professor at the University of California, Berkeley, had
started lobbying for his state to protect carnivorous mammals in 1912.
He and his colleagues were ferocious critics of the Bureau of Biological Survey, whose extensive predator extermination programs, carried
out in the name of game and livestock protection, would later draw the
ire of Rosalie Edge. "It is a curious perversion," Grinnell wrote, "when
'conservation' is appealed to to justify *destruction*."

In 1922, Leopold submitted an article to *The Condor*, a journal helmed
by Grinnell, in which he described his discovery that roadrunners prey
on quail—a favorite quarry of Leopold's. "I have never before killed a
road-runner," he wrote, "but they are now on my 'blacklist' and will stay
there until somebody proves that this was an exception to their usual
habits." The journal's managing editor responded that in his opinion,
the roadrunner was just as desirable a bird as the quail, adding somewhat snarkily, "I suppose, though, that you look at the matter from the
'game protective' standpoint." Leopold admitted that he had been a "bit
hasty," and assured the editor that he was "heartily in sympathy" with
Grinnell. The line was excised.

Leopold began to seriously consider the consequences of predator
eradication in the mid-1920s, after the deer on the Kaibab Plateau,
north of the Grand Canyon, became so numerous in the absence of
mountain lions, wolves, and other predators that they munched down
all the edible shrubs and trees. Over several winters, thousands of
deer starved to death, causing national consternation. Novelist Zane
Grey, perhaps inspired by one of his own westerns, even led an aborted
attempt to drive some of the deer into the Grand Canyon and across
the Colorado River. Panicked officials issued deer-hunting permits
three to a person, and the herd was reduced by force. (The disaster on
the Kaibab has been examined and reexamined, and while the most
recent analysis suggests that predator eradication did indeed contrib-

ute to the deer population explosion, researchers caution that uncertainty remains.)

Leopold, who was already working in Madison, came to suspect that
in such cases, the biotic pyramid was missing its peak. In 1929, when a
deer herd in the Gila National Forest grew large enough to threaten its
food supply, he quietly suggested to a colleague that "letting the lions
alone for a while" might head off a crisis. Over the next several years,
influenced by developments in ecology and his own findings during the
game survey, Leopold continued to shift his position on predators. By
the early 1930s, he had begun to argue publicly against predator trapping and poisoning, telling an audience at the Nineteenth American
Game Conference that predator control should be employed "reluctantly,
selectively, and only after other measures fail to restore the game."

In 1935, when the National Rifle Association magazine published
an article that praised eagle hunting as "the purest of all rifle sports,"
Leopold wrote to the organization's president with a bitter invitation
for the author: "I would infinitely rather that Mr. Burch shoot the vases
off my mantelpiece than the eagles out of my Alaska." Leopold, diplomatic by nature, had less appetite for confrontation than Rosalie Edge.
"I am in general opposed to the poisoning campaign," he wrote to her
after receiving an especially hotly worded Emergency Conservation
Committee pamphlet, "but there is nothing to be gained by injustice to
those who think otherwise." (She cheerfully replied that "those who tell
the truth forcefully must not hope for commendation from all.") Their
sentiments had converged, however, and his concern, like hers, was for
the entire biotic pyramid.

Less than a decade later, in 1943, Leopold attempted to put his principles into practice as a Wisconsin state conservation commissioner.
Faced with a severe overpopulation of deer in the forests of northern
Wisconsin, Leopold and most of his fellow commissioners recommended a hunting season that allowed the killing of does and young
deer, surmising that since the state had nearly exterminated its wolves,
human hunters needed to fill the vacant role of top predator. Public

opposition was passionate—critics leveraged the Disney movie *Bambi*, which had been released the previous summer—and state legislators backed away from the policy. Instead, they reinstated a bounty on the state's remaining wolves, resulting in hungry deer herds that chewed up hundreds of millions of young trees. The commissioners were vilified, and Leopold reflected ruefully that "the public, unlike the mule, kicks both fore and aft."

During this bruising experience, Leopold was writing and revising the essays that would eventually make up *A Sand County Almanac*. He sent several of his drafts to Albert Hochbaum, a former graduate student of his who was managing a duck station in Manitoba. Hochbaum, a talented artist, was contributing pen-and-ink drawings to the book, but Leopold asked for his editorial comments, too, and he got them.

"The lesson you wish to put across is the lesson that must be taught—preservation of the natural," Hochbaum wrote to Leopold after reading several drafts. "Yet it is not easily taught if you put yourself above other men." The essays would be more effective, Hochbaum said, if Leopold could bring himself to expose his own errors—if he "let fall a hint that in the process of reaching the end result of your thinking you have sometimes followed trails like anyone else that lead you up the wrong alleys." Hochbaum suggested that Leopold had some atoning to do, especially concerning predators. "I just read they killed the last lobo in Montana last year," he wrote. "I think you'll have to admit you've got at least a drop of its blood on your hands."

Leopold was intrigued but wary. "Your point is obviously well taken, and I think I can see several opportunities for admitting specifically that we all go through the wringer at one time or another," he responded. "The question is how to do this without spoiling literary effects." Hochbaum, perhaps sensing that literary effects were not Leopold's only concern, kept after his mentor, and eventually Leopold took his advice. The result was "Thinking Like a Mountain," one of Leopold's best-known essays.

In the essay, Leopold compresses his years-long conversion into a single event during the turbulent summer of 1909, when he was irritating his Forest Service surveying crew with his youthful know-it-allism. One day, he and a companion decided to take their lunch on the edge of a river canyon. Far below their feet, they saw what they believed to be a doe breasting through the water. When the animal emerged, they realized she was a female wolf, accompanied by a half dozen nearly grown pups. "In those days, we had never heard of passing up a chance to kill a wolf," Leopold wrote. Without much thought, he and his companion raised their rifles and began firing into the canyon. When they descended to the riverbank, the adult wolf was wounded but still alive.

Leopold, in his telling, watched what he called "a fierce green fire" fade from her eyes. "I was young then, and full of trigger-itch; I thought that because fewer wolves meant more deer, that no wolves would mean hunters' paradise," he wrote. "But after seeing the green fire die, I sensed that neither the wolf nor the mountain agreed with such a view."

Just south of the Leopold shack lies the Baraboo Range, an irregular ring of quartzite hills that formed well over a billion years ago, was preserved under a layer of marine sediments, and has since been exposed by erosion. Today, it is one of the few places on earth where it's possible to walk on a surface shaped in the late Proterozoic, before organisms with skeletons appeared on the planet. John Muir, who spent most of his childhood on a farm nearby, used to escape from chores and his tyrannical father by climbing an ancient rhyolite knob called Observatory Hill. From the top, Muir could almost see the river bottoms that Leopold would one day plant with pines and prairie sod.

The neighborhood's living inhabitants have changed since Muir, and later Leopold, walked here. Hundreds of the pines that Leopold and his family planted are still alive, and tower over the shack. In 2015 the Aldo Leopold Foundation, which owns the land, began to thin some of the floodplain forests in order to restore grassland bird habitat. Leopold's great-grandchildren come to visit, and several of his descendants are active in the foundation, whose mission is not only to maintain Leopold's legacy but to continue the restoration work he began.

On April 21, 1948, Aldo, Estella, and Estella Jr. were at the shack with their new dog, a German shorthaired pointer named Flicky. After breakfast, Aldo noticed a plume of smoke rising in the distance, near a neighbor's farmhouse. The wind was blowing toward the shack, and he became uncharacteristically nervous—"really jittery," Estella Jr. remembered. He pulled a backpack pump down from the rafters and filled it with water, then headed off toward the smoke on foot. Estella Jr. called the fire department, and she and her mother stood guard near the shack with rakes and wet brooms, ready to put out stray embers. After about an hour, the smoke seemed to lessen, and Estella Jr. saw a man she didn't know walking toward the shack. "Your father is not well," he told her.

She looked at his face. "Oh, it is worse than that," she said. "Isn't it?"

While walking the perimeter of the blaze, Leopold had suffered a heart attack. He set down the pump and lay down on his back, hands folded over his chest and head pillowed by a clump of grass. In his pockets were two University of Wisconsin faculty cards, a university club card, a driver's license, a photograph of Estella, and a tiny ruled notebook in which he had penciled the morning's observations: one blue heron, three pileated woodpeckers, bloodroot in bloom. The fire had scorched the pages.

Just over a month later, Estella, Estella Jr., and Nina returned to the shack. "We did not know what to do with ourselves except to continue doing what Dad would have done," remembered Estella Jr. She noted which flowers were blooming—blue oxalis, white dodecatheon, crowfoot violet—and which birds were singing: quail, woodcock, red-headed woodpecker. A hummingbird buzzed through the yard.

"Those plants were going to continue blooming and the birds calling without Dad watching over them," Estella Jr. reflected. "How could *that* be?"

Leopold had planned to spend the summer finishing his essay collection. In the months after his death, Estella, their son Luna, and a small group of friends and colleagues took over the task, making only a few substantive changes to the manuscript. *A Sand County Almanac and Sketches Here and There* was published in the fall of 1949 to enthusiastic reviews, and gained a devoted readership in conservation circles. For more than two decades, though, it remained something of a cult classic; Susan Flader, Leopold's first biographer, was introduced to the book by her college German professor and for years felt that it was a "private discovery." Not until 1970, when the book was published in a mass-market edition to coincide with the first Earth Day, did it become the widely quoted work it is today. (My ragged paperback copy, published in 1991, is from its thirty-first printing.)

The book's most famous essay, "The Land Ethic," is a distillation of Leopold's lifetime of thinking about conservation. Since his Forest Service days, he had understood that the conservation movement had to counter humans' "philosophical imperialism" toward other species and their habitats, especially on private land. Doing so required more than what he dismissed during his university arboretum address as "some fish and game laws"; it required more than economic incentives and reforms, though he believed those were necessary as well. The "only visible remedy," he concluded in his essay, was the development of an "ethical obligation on the part of the private owner"—the kind of obligation that had led to his own father's self-imposed hunting limits. Cautioning that "nothing so important as an ethic is ever 'written,'" he nevertheless put his into words: "A thing is right when it tends to preserve the integrity, stability, and beauty of the biotic community. It is wrong when it tends otherwise."

Leopold was not a trained philosopher, and over the decades his ethic has been variously criticized for being too vague or too prescriptive, too capitalistic or too communitarian. His vision of conservation as practiced by communities of small landowners did not foresee that, during the second half of the twentieth century, agricultural mechanization and market pressures would lead to widespread land consolidation on the plains, leaving the region with fewer, larger farms.

But the uses of a land ethic are not limited to landowners. While the twenty-five words of the ethic have been scrutinized by generations of scholars, I find myself returning at least as often to another paragraph in the essay—to Leopold's description of his ethic's hoped-for effects. "In short, a land ethic changes the role of *Homo sapiens* from conqueror of the land-community to plain member and citizen of it," he wrote. "It implies respect for his fellow-members, and also respect for the community as such."

Plain member and citizen. Respect for one's fellow members, and respect for the community. Plain members and citizens of what Leopold called the land-community are not only different from conquerors but different from the Christian concept of stewards, put into Eden "to dress it and to keep it." The closer biblical parallel, perhaps, is the commandment to love thy neighbor as thyself; neighbors care for one another, in mutual relationships, while allowing one another as much autonomy as possible within the bounds of their community. Though human relationships with the members of other species range from domestic to predatory, benevolent to profit-driven, few are entirely one-way, and almost all have the potential for some degree of autonomy. The role of plain member and citizen, it seems to me, allows all of us to acknowledge our interdependence with the rest of life—as well as our species' unique ability, and responsibility, to guard its independence from ourselves.

All five of the Leopold children became accomplished scientists and conservationists, and all, at some point, acquired their own shacks on pieces of run-down land. The only surviving sibling is Estella Jr., an expert in fossil pollen whose research has taken her to China and the South Pacific. At ninety-three, she maintains her laboratory at the University of Washington in Seattle, where she taught for decades, and keeps tabs on her shack in Colorado, which is cared for by a niece and nephew. The relationship between Leopolds and their shacks, she told me, is mutual: "We restored those places. And they restored us."

• • • •

One month after giving his 1934 address at the university arboretum, when the skies were still gritty with Oklahoma dust, Leopold, his son Starker, and Leopold's brother Carl took a fishing trip to northern Wisconsin. On their way home, they stopped near Buffalo Lake to watch a pair of sandhill cranes, one of an estimated twenty breeding pairs left in the state. As the men scanned the marsh with binoculars, the birds suddenly flew up from the edge of the nearby woods, trumpeting loudly. "It was a noble sight," Leopold noted in his journal. The day sparked a lasting fascination with the ungainly elegance of cranes, and led to his essay "Marshland Elegy": "His tribe, we now know, stems out of the remote Eocene," Leopold wrote. "When we hear his call we hear no mere bird. We hear the trumpet in the orchestra of evolution. He is the symbol of our untamable past, of that incredible sweep of millennia which underlies and conditions the daily affairs of birds and men."

Sandhill crane populations had been pummeled by commercial and recreational hunters, and in 1937, when Leopold published his essay, an elegy for the species must have seemed all too appropriate. But the hunting restrictions championed decades earlier by Audubon activists were taking hold, and slowly, sandhill numbers were beginning to rise.

Meanwhile, the federal government—taking advantage of the opportunity Leopold and his colleagues had alluded to in their game policy—had begun to buy marginal farmland for restoration as wildlife habitat. Over the next twenty years, some fourteen million acres of wildlife refuge were added to the national system, including land within the sandhills' historic range in Wisconsin. On public and private land, wetlands that had been drained and converted to farmland began to be restored. By the mid-1960s, annual bird counts by Audubon Society volunteers and studies by government biologists were documenting more than four hundred sandhill cranes in the eastern United States. By the 2010s, they were counting more than forty-five thousand.

Leopold left no clear record of seeing cranes at the shack, but in the 1970s, Estella and Estella Jr. spotted a pair landing in a marsh on the property. In the year 2000, a second pair nested in another nearby wetland. On a December evening in 2019, I stood just upstream from the shack and watched as thousands of cranes croaked and rattled overhead, all winging toward the river after a day of foraging for grain in the surrounding fields. The cranes pause by the shack for several weeks each fall, fattening themselves up for the grueling migration to come. Within a few days these birds would head south, flying high over Chicago and then on to northern Indiana. As the birds whose ancestors Leopold had eulogized wheeled low over the Wisconsin River, legs dangling in preparation for landing, their feathers glowed fire-orange in the setting sun. It was a noble sight indeed.

Sandhill cranes are only one of the world's fifteen species of cranes, and representatives of all fifteen live on the grounds of the International Crane Foundation, located just five miles from Leopold's shack. The foundation is home to black crowned and grey crowned cranes from Africa, with their coronas of spiky golden feathers; sarus cranes, the tallest flying birds in the world, from southern Asia and Australia; and Siberian cranes, which are severely threatened in China by wetland development and poaching—the same factors that nearly doomed the sandhill crane a century ago. The first time I visited, a towering sarus crane was stooped protectively over her newborn chick, purring like a cat.

The foundation, started in 1973 by two young Cornell University graduates, works with partners in China, Cambodia, India, South Africa, Zambia, and elsewhere to protect and restore crane habitat and reintroduce cranes. Its international outlook owes much to Leopold, who in his later years warned against provincialism in conservation. "The impending industrialization of the world," he wrote in 1944, "means that many conservation problems heretofore local will shortly become global." But the foundation's work might not have been possible at all without the foresight of a British biologist named Julian Huxley, who helped unite species conservation efforts into an international movement.

Chapter Five

THE PROFESSOR AND
THE ELIXIR OF LIFE

"Dear Grandpater," four-year-old Julian Huxley wrote in 1892, "Have you seen a Water-baby? Did you put it in a bottle? Did it wonder if it could get out? Can I see it someday?"

Julian's grandpater was the renowned British zoologist Thomas Henry Huxley, known as T. H., and Julian's letter was inspired by the peculiar children's classic *The Water-Babies*, a fairy tale that doubles as an evolutionary fable. The elder Huxley was an early defender of Charles Darwin's theory of evolution by natural selection, and he appears in *The Water-Babies*, in caricature, alongside several of his contemporaries. "They are very wise men; and you must listen respectfully to all they say," author Charles Kingsley told his young readers. "But even if they should say, which I am sure they never would, 'That cannot exist. That is contrary to nature,' you must wait a little, and see; for perhaps even they may be wrong."

Young Julian was apparently unsurprised by his own grandfather's appearance in one of his favorite books, but he was curious about the amphibious humans the book described. The elder Huxley was known as a fierce public debater—he was nicknamed "Darwin's bulldog"—but he had a soft spot for his oldest grandson, and he answered Julian's inquiry with tenderness and respect:

My dear Julian—I never could make sure about that Water Baby.

I have seen Babies in water and Babies in bottles; but the Baby in the water was not in a bottle and the Baby in the bottle was not in water. My friend who wrote the story of the Water Baby was a very kind man and very clever. Perhaps he thought I could see as much in the water as he did—There are some people who see a great deal and some who see very little in the same things.

When you grow up I dare say you will be one of the great-deal seers, and see things more wonderful than Water Babies where other folks can see nothing.

Julian, who was already captivated by plants and animals, may well have read and remembered another story from *The Water-Babies*: the melancholy encounter between the main character, Tom, and a large black-and-white bird standing alone on a rock in the North Atlantic. "And a very grand old lady she was, full three feet high, and bolt upright, like some old Highland chieftainess," Kingsley wrote.

"I never could make sure about that Water Baby": Richard Owen and Thomas Henry Huxley in an 1885 illustration for *The Water-Babies*.

The bird, which Kingsley called the Gairfowl, was also known as the great auk. "Once we were a great nation, and spread over all the Northern Isles," the Gairfowl tells Tom. "But men shot us so, and knocked us on the head, and took our eggs—why, if you will believe it, they say that on the coast of Labrador the sailors used to lay a plank from the rock on board the thing called their ship, and drive us along the plank by hundreds, till we tumbled down into the ship's waist in heaps; and then, I suppose, they ate us, the nasty fellows!"

Kingsley's book was first published in 1863, two years before ornithologist Alfred Newton mournfully declared the great auk to be "a thing of the past." When the younger Huxley met the water-babies, the great auk had not been seen for four decades.

Julian Huxley would follow his early curiosity about other species into a career so varied that biographers have struggled to capture it. He was a scientist, a public intellectual, and an international statesman, sometimes all at once. He was also, in many ways, an old-fashioned elitist—"a Victorian thinker fated to live in an unsympathetic modern age," as one scholar puts it. As his grandfather predicted, though, he was one of the great-deal seers, and he saw that the future of conservation lay beyond borders.

Julian Huxley was only six months younger than Aldo Leopold—he was born in the summer of 1887, during the celebrations of Queen Victoria's fiftieth year on the throne—but he grew up in a different world: not on the North American frontier but in the center of the still-expanding British Empire, into a family already absorbed in the study of life.

His grandfather Huxley, the son of a schoolmaster, had battled his way into England's intellectual elite, wearing down class barriers with smarts, charm, and a fearsome work ethic. At twenty-one, newly graduated from medical school, T. H. Huxley secured himself a place in the Royal Navy, and in December of 1846 he sailed to the South Pacific as an assistant surgeon on the aging frigate HMS *Rattlesnake*. During the nearly four-year voyage, the crew surveyed the east coast of Australia, the Torres Strait, and the southern coast of New Guinea. When Huxley's doctoring duties permitted, he practiced natural history, studying mollusks and anemones in his tiny cabin through a microscope lashed to a bench. He marveled at the unfamiliar world around him; when the *Rattlesnake* sailed through a colony of phosphorescent tunicates, he recalled, "she drifted hour after hour, through this shoal of miniature pillars of fire gleaming out of the dark sea, with an ever-waning, ever-brightening, soft bluish light, as far as the eye could reach on every side."

Back in London, Huxley set out to establish a scientific career, no easy task at a time when most "men of science" financed their work with family wealth or employment in other fields. Huxley demanded and eventually won respect for his research—on the *Rattlesnake*, he had cleared up a point of confusion in invertebrate taxonomy, and he would go on to do pioneering work in animal physiology—but he found that his greatest genius was for communication. He cut a compelling figure at the lectern, with dark hair that he wore long for the time and an ability to exploit what family biographer Ronald Clark breathlessly describes as "his grave, earnest expression, his dark, almost Svengali-like eyes and his full mobile lips." The science fiction writer H. G. Wells, who was a student when he heard Huxley lecture, remembered decades later that "only afterwards did you realize how much that quiet leisurely voice had said and how swiftly it had covered the ground." Huxley made a point of speaking to working-class audiences, and he soon gained a popular following; it was said that cab drivers forgave his fares, and messengers saved his envelopes in order to show his handwriting to their families.

In December 1859, *The Times* of London sent Huxley a copy of a new book by Charles Darwin, requesting that he review it at length. Huxley knew Darwin as a colleague and friend, and was aware that for more than twenty-five years, ever since Darwin's own stint at sea on HMS *Beagle*, he had been cogitating about how species formed and whether and how they changed over time. While other naturalists—including Darwin's grandfather, Erasmus—had wondered if new species could arise from old ones, most Europeans and North Americans still believed that species were static creations of God. And few doubted that humans were a special species, set over all the others.

Darwin's firsthand observations of birds in the Galápagos Islands, and of the pigeons and cabbages he raised at his home in England, convinced him that these beliefs were false. Small differences among individuals, he realized, allow some members of a population to survive longer and produce more offspring, bequeathing their advantages to the next generation. Over hundreds or thousands of generations, this process of selection and inheritance leads to new kinds of life which, as he put it, strive in their turn to "seize on each place in the economy of nature."

Darwin, a wealthy "man of science," was cautious and retiring by temperament, and he kept his explosive insight to himself for years. "I am almost convinced (quite contrary to the opinion I started with) that species are not (it is like confessing a murder) immutable," he tremblingly confided to his close friend, the eminent botanist Joseph Hooker, in 1844. Not until the young explorer Alfred Russel Wallace sent Darwin a draft of his own alarmingly similar theory was Darwin spurred to publish. Darwin's theory of evolution was introduced with Wallace's at a meeting of the Linnaean Society in 1858, and Darwin's new book, *On the Origin of Species*, was his first public elaboration of it.

T. H. Huxley was delighted by the book, describing it later as a "flash of light, which to a man who has lost himself in a dark night, suddenly reveals a road which, whether it takes him straight home or not, certainly goes his way." (His very first reaction, as he remembered it, was less poetic: "How extremely stupid not to have thought of that!")

On the Origin of Species made only coy references to human evolution—"Light will be thrown on the origin of man and his history," Darwin wrote—but Huxley immediately understood that the book challenged the Christian belief in human dominion and would need a stout defender. "I trust you will not allow yourself to be in any way disgusted or annoyed by the considerable abuse and misrepresentation which, unless I greatly mistake, is in store for you," he wrote to Darwin, adding, "I am sharpening up my claws and beak in readiness."

The book did cause a furor, and Huxley duly unsheathed his claws. He struck his most famous blow on Darwin's behalf in the summer of 1860, when he debated the bishop Samuel Wilberforce at Oxford University. (Darwin himself was suffering from stomach pain and sent his regrets.) Wilberforce, twenty years older than Huxley, was renowned for his skilled oratory—his critics called him "Soapy Sam"—and several hundred men and women showed up to hear him smoothly dismiss what he had called "our unsuspected cousinship with the mushrooms." Of course species were fixed entities, he said. Of course humans were special.

At the end of the bishop's address—according to the most detailed of the surviving eyewitness accounts—he turned to Huxley and "asked if he had any particular predilection for a monkey ancestry, and, if so, on which side—whether he would prefer an ape for his grandfather, and a woman for his grandmother, or a man for his grandfather, and an ape for his grandmother"?

Such a personal attack—one implicating the ladies, no less!—was considered highly improper, and Huxley knew that Wilberforce, for all his diplomatic acuity, had handed him an opportunity. As Huxley later recounted to a friend, "If then, said I, the question is put to me would I rather have a miserable ape for a grandfather or a man highly endowed by nature and possessed of great means and influence and yet who employs these faculties and that influence for the mere purpose of introducing ridicule into a grave scientific discussion—I unhesitatingly affirm my preference for the ape." This retort, Huxley reported with satisfaction, was met with "unextinguishable laughter among the people."

Huxley and his allies went on to make their case for evolution, and left Oxford convinced of victory. "I see daily more & more plainly," a grateful Darwin wrote to Huxley soon afterward, "that my unaided Book would have done *absolutely* nothing."

Darwin's bulldog had scored a decisive rhetorical point. Still, neither Huxley nor evolution itself could answer Wilberforce's deeper question: If humans were not divinely distinct from the rest of life, what exactly were they? Darwin left his fellow Victorians in limbo, dangling inelegantly between mushrooms and minor gods, and many never forgave him or Huxley for it. In 1895, when Huxley was on his deathbed, one of his daughters had to intercept the steady stream of letters from strangers who wrote to say how happy they were that he was going to hell.

Huxley himself was more intrigued than bothered by this existential uncertainty. Long a religious skeptic, he lost his faith for good after his eldest son died of scarlet fever at age three, and he later coined the word "agnostic" to describe his attitude toward the divine. He would follow his reason only as far as it could take him, he said, and since science could neither prove the existence nor the absence of God, he would remain uncommitted. "The question of questions for mankind," he once mused, "the problem which underlies all others, and is more deeply interesting than any other, is the ascertainment of the place which Man occupies in nature and of his relations to the universe of things."

The place of *Homo sapiens* among the rest of life was complicated not only by Darwin's theory of evolution but by his quiet acknowledgment that other species "have been exterminated, either locally or wholly, through man's agency." Despite Darwin's assurance that extinction was nothing to marvel at, human-caused extinction fired the Victorian imagination; two years after Charles Kingsley memorialized the great auk in *The Water-Babies*, Lewis Carroll caricatured himself as a dodo in *Alice's Adventures in Wonderland*. Naturalists began a frenzied search for dodo remains, a species that by that time had been extinct for nearly two centuries—so long that some people doubted it had existed at all.

Humans were not as powerful as they had believed, but they were powerful enough to make other species disappear. T. H. Huxley had helped administer the first shock, and his grandson Julian would face the rippling effects of both—at home in England, and around the world.

When Julian was four years old, he defied his grandfather's order to stay off a wet lawn. T. H., amused, recognized a fellow insurgent. "I like that chap!" he said. "I like the way he looks you straight in the face and disobeys you!"

Julian inherited what were by then the classic Huxley characteristics: a birdlike frame, an intensely inquisitive mind, a disregard for authority, and a short fuse. (When Julian was in his fifties, *Life* magazine would describe him as a "bespectacled, bony-faced scholar who looks placid but erupts in fits of temper or kinetic conversation.") Though T. H. was still the most famous Huxley, most of his descendants were both distinguished and distinctive, and clan membership came with expectations. "You are marrying into one of the great atheist families," an elderly aunt once said to a prospective in-law. "I know you are an atheist now, but will you be able to keep it up until you die?"

In the summer of 1909, after finishing his undergraduate studies in zoology at Oxford, Julian attended a celebration of the centenary of Darwin's birth and the fiftieth anniversary of the publication of *On the Origin of Species*. Sitting side by side on the dais were his grandmother Henrietta Huxley, who at eighty-four still read each week's edition of the journal *Nature*, and Joseph Hooker, who had been one of T. H.'s closest allies in Darwin's defense. As Julian listened to the addresses, as he later recalled, "I realized more fully than ever that Darwin's theory of evolution had emerged as one of the great liberating concepts of science, freeing man from cramping myths and dogma." All of his own work, he resolved, "would be undertaken in a Darwinian spirit." In the decades to come, evolution would not only guide his research but inform his philosophy of life.

His path into adulthood would not be easy. Julian's mostly idyllic youth had come to an end the previous year, when his mother, Julia, died of cancer at forty-six. "It is very hard to leave you all," she wrote in a farewell message to Julian, "but after these weeks of quiet thought, I know that all life is one—and that I am only going into another room." Julian was devastated, and his youngest brother, Aldous, then fourteen, even more so. Aldous's suffering was soon compounded by an eye infection, which blinded him for eighteen months and left him severely nearsighted. Forced to abandon his plans to study medicine, he turned to literature, beginning the work that would make him the best-known Huxley of his generation: his dystopian novel *Brave New World*, published in 1932, was a bestseller in its time and is considered a classic today.

After leaving Oxford, Julian cast about for direction. He spent a year at the Zoological Station in Naples, Italy, one of the world's oldest marine research laboratories. He applied, unsuccessfully, for a place on Robert Scott's disastrous expedition to the South Pole. He had returned to Oxford and a poorly paid lectureship when, in the summer of 1912, he was offered a position as chairman of the biology department at the new Rice Institute in Texas.

It was a thrilling opportunity for a young biologist, and he almost missed it. In 1913, after the drawn-out failure of a love affair, he was struck by the depression that ran in the Huxley family, and was hospitalized for a time. In 1914, after a year at Rice, he was again overcome and retreated to England to recover, this time at the same hospital where his younger brother Trev was also being treated for depression. Only days after Julian improved and left the hospital, Trev hanged himself in the nearby woods. His death would weigh heavily on his surviving siblings for the rest of their lives.

Heartsick, Julian returned to Texas, where he found solace in other species. He spotted his first hummingbird, hovering over a spider-lily in Houston, and his first scissor-tailed flycatcher, gray and pink with an elegant forked tail. During a university break, he visited Avery Island in Louisiana, where Tabasco pepper sauce baron Edward McIlhenny had established a private nature reserve and introduced about a dozen snowy egrets. The island, a former sugar plantation, had since attracted thousands of birds, and Julian withstood the mosquitoes long enough to make some observations of bird behavior. He noted that the egrets engaged in mutual courtship displays, with both members of a pair showing off their gossamer fans of feathers—the same feathers that were still commanding high prices as hat ornaments.

Whenever he could, Julian drove his Model T Ford ("a gallant little machine") across the Texas plains, marveling at the scale of North America. His time in the United States made him feel, he said, that "all the familiar institutions and ideas of my own country were not the inevitable and permanent things that they had seemed."

As Julian awakened to the wider world, European and North American conservationists were also expanding their ambitions, seeking to protect species beyond their own continents. Most of the first specifically international conservation efforts were extensions of colonial power, enacted to protect economic interests overseas. In southern Africa, the Dutch instituted game protection laws in 1657, and the British imposed a succession of hunting regulations in the 1800s.

Like William Hornaday in Montana Territory, many European conservationists in Africa believed that subsistence hunters were to blame for declining game numbers. Others argued that responsibility lay with their fellow colonists. None could deny, however, that the arrival of European guns and railways had intensified commercial hunting, especially that which supplied the immensely lucrative trade in elephant ivory. When Britain and other colonial nations reached into eastern and central Africa in the 1880s, so did ivory hunters, and despite African rulers' efforts to limit the slaughter, elephants were killed by the thousands. In 1900, concerned about the future of the ivory trade, European powers signed an agreement to protect the African animal species "which are either useful to man or are harmless." (The convention actually encouraged the killing of hyenas, lions, otters, and other species that the signatories considered to be pests.)

Though the convention never came into force, it spurred conservation-minded Europeans to more substantive action. In 1903, Edward North Buxton, a British aristocrat and big-game hunter who had fought for public access to forests at home in England, organized social and scientific notables in support of the goals of the 1900 convention. The group called itself the Society for the Preservation of the Wild Fauna of the Empire, and much like the Boone and Crockett Club, it used its collective political influence to protect the species its members loved to hunt. Though the British press lampooned the society as a group of "penitent butchers"—a nickname its members resisted at first, but eventually adopted—the group outlasted the ridicule, and remains active as the more humbly named Fauna & Flora International.

Meanwhile, European bird protection societies had begun to call for international species conservation measures closer to home. The first such treaty, signed by several European countries in 1902, was limited to "useful birds," and, like the 1900 convention, its primary accomplishment was to rouse complacent European conservationists. Eight years later, the Swiss conservationist Paul Sarasin outlined his vision for *Weltnaturschutz*—global conservation—at an international zoology conference in Austria. "All the wild, living higher fauna of our planet is destined to complete destruction if all those who are capable of realizing the danger do not oppose it with their last energy," he said.

Sarasin, the son of a wealthy industrialist, was already known throughout Europe for his scientific adventures in Southeast Asia, and he convinced the Swiss government to host an international conservation congress in 1913. Representatives from seventeen countries attended, and agreed to take preliminary steps toward the formation of an intergovernmental conservation organization. But just as Sarasin was issuing invitations to a follow-up meeting, the outbreak of the First World War brought the project to a sudden end.

Huxley stayed in Texas until 1916, when, feeling obligated to join the war effort, he returned to England to serve in the Intelligence Corps. Shortly before deploying to Italy, he visited Aldous at the home of Lady

Ottoline Morrell, a central figure in the Bloomsbury circle of writers and artists. There he met Lady Ottoline's young Swiss governess, Juliette Baillot, who had come to England after a sheltered childhood and was both shocked and intrigued by the bohemian Bloomsbury set. Her first impression of Julian, recalled in her memoir decades later, was less than flattering: "He lacked the pensive beauty of Aldous, his face immature, his head definitely dolichocephalic." The two began to correspond, however, and their letters became confiding. Juliette spoke of her loneliness in England, and Julian of his breakdown and recovery. Julian proposed almost immediately upon returning home, and they were married in 1919.

Theirs would not be a tranquil marriage. Julian continued to suffer depressive spells, and his consuming ambition often made him, as Juliette put it, "impatient of intrusion or contradiction and of the very existence of others." Julian had several affairs, including a brief romance with the poet May Sarton—who promptly fell in love with Juliette, beginning a passionate correspondence that lasted for decades. But Julian and Juliette shared a curiosity about the world, and Julian, at his best, was an extraordinary companion in it: "Julian had the gift of seeing with the inner eye, of discovering hidden treasures in grass and tree, of knowing the name of everything," Juliette remembered. "Plants and birds, rivulets and villages, they became one's own because he knew them."

The couple moved to Oxford, where Huxley, after being interrupted by another breakdown, continued his studies of animal development and behavior. Within a few months, he had observed that hormones extracted from a mammalian thyroid gland could induce metamorphosis in the axolotl, a Mexican salamander that typically reaches adulthood without abandoning its larval form. When he reported his results in a letter to *Nature*, in January 1920, the press hurried to anoint a new Huxley genius, announcing that "one of the most brilliant of our young biologists" had discovered the "Elixir of Life."

Huxley had, of course, done no such thing—and as he would later

Juliette and Julian Huxley, photographed by Lady Ottoline Morrell in 1924.

learn, he was not even the first to perform the metamorphosis experiment. When he wrote a column for the *Daily Mail* to correct the record, he discovered that he had the family knack for explaining science to the public, and over the next few years he began to publish popular essays. Many of Huxley's colleagues already considered him a bit too versatile for his own good, and the addition of a second career exasperated them further. "For goodness sake do decide which branch of biology you are expert in," Cambridge zoologist George Bidder wrote to him in 1925.

Like his grandfather, Huxley eventually moved away from research in favor of translating its findings for a broader audience. In 1926, he began a tempestuous collaboration with H. G. Wells, producing an encyclopedic but hugely popular series called *The Science of Life*. Julian and Juliette then spent three months in Kenya and Uganda, where Julian surveyed the state of colonial science education on behalf of the British government. On the Ugandan border, the Huxleys climbed the slopes of the Virunga volcanoes, passing through stands of giant parsley and curtains of dangling orchids as they looked for signs of mountain gorillas.

Back in England, Huxley directed a short documentary about a colony of gannets on a rocky island off the coast of Wales, a film which won an Academy Award. In 1935, he was hired as the director of the London Zoo, which he aimed to transform into "the centre and focus of popular interest in every aspect of animals and animal life." The Huxleys lived on zoo grounds, and their elder son, Anthony, became, as Huxley family biographer Ronald Clark recounts, "the only horseman ever unseated in England by a charging rhinoceros." When the Second World War broke out, the zoo's venomous snakes and spiders were killed to prevent their escape, but many of the large animals remained in London—and Huxley continued to spend much of his time in the city, too. The zoo was bombed during the Blitz, and at one point Huxley found himself chasing a zebra through the blacked-out city streets. When he later confessed to a zookeeper that he'd been afraid to approach the cornered animal, the keeper responded, "Oh, you needn't have worried, sir. He's a biter, not a kicker."

Huxley found time to study the response of the animals to the bombing raids, noting that they appeared far less disturbed than might have been expected. His tenure at the zoo was controversial, however—some board members were unimpressed by his research and public education initiatives, preferring to focus on "the display of our Menagerie"—

and while he was on a speaking tour of the United States in 1942, his position was eliminated.

During these years, Huxley was also a regular guest on *The Brains Trust*, a radio program that became a touchstone of wartime life in Britain. Huxley and other scholars offered answers to questions sent in by listeners, including servicemen and women: How does a fly land on the ceiling? How does a kangaroo clean out its pouch? One group of schoolgirls asked why it was necessary to learn algebra; a teacher asked for the panel's definition of hate. When a listener from Glasgow asked which character from fiction the panel would most like to see on the screen, Huxley named Don Quixote.

When he was in his eighties, Huxley wondered at his lifelong restlessness. "I seem to have been possessed by a demon," he wrote, "driving me into every sort of activity, and impatient to finish anything I had begun." His seemingly fragmentary career, however, was guided by his grandfather's "question of questions" about the place of humans in relation to the rest of life—and by an ambition to not only answer the question, but change the answer.

Toward the end of his life, T. H. Huxley had brandished his well-sharpened beak and claws at those who proposed to meddle with the course of human evolution. There was, he wrote, "no hope that mere human beings will ever possess enough intelligence to select the fittest." Though Julian loved and admired his grandfather, on this point he looked T. H. straight in the face and disobeyed him.

The phrase "survival of the fittest" was coined not by Darwin but by British philosopher Herbert Spencer, his slightly younger contemporary. Both Spencer and Darwin were deeply influenced by the cleric and scholar Thomas Robert Malthus, who in the late 1700s had observed the "constant tendency in all animated life to increase beyond the nourishment prepared for it." Malthus argued that human populations, like populations of many other species, increased exponentially when unchecked—and that human numbers, barring what Malthus called "preventive checks" on reproduction, were checked by "misery," usually in the form of famine.

Malthus's view of life as a competition for limited resources led Spencer to argue that the suffering of the "worst fitted" to society would lead to the "ultimate perfection" of humanity. "The poverty of the incapable, the distresses that come upon the imprudent, the starvation of the idle, and those shoulderings aside of the weak by the strong . . . are the decrees of a large, far-seeing benevolence," Spencer wrote in the mid-1800s.

Malthus's ideas also underlay Darwin's theory of evolution, but Darwin understood that the process of natural selection—in any population, human or otherwise—was neither finite nor destined to improve humanity. (In the 1990s, paleontologist Stephen Jay Gould memorably compared evolution to "a drunkard's walk.") Spencer's vision of a society perfected by selection nonetheless became known as "social Darwinism," and variations of it were espoused by some Darwin adherents. German zoologist Ernst Haeckel, who had coined the term "ecology," sorted humanity into twelve species with Caucasians "as the most highly developed and perfect." He believed that these "species" were competing with one another for survival—and, as in his artistic renderings of the evolutionary tree, for a place at the supposed pinnacle of life. With the rise of German nationalism, Haeckel came to believe that nations, too, were battling for dominance, fit against unfit.

In the late 1800s, Darwin's cousin Francis Galton, whose scientific pursuits ranged from geography to meteorology to forensics, became convinced that by manipulating human reproduction—by controlling who reproduced, and with whom—humans could speed up the perfecting process. "What nature does blindly, slowly, and ruthlessly, man may do providently, quickly, and kindly," Galton wrote. "The improvement of our stock seems to me one of the highest objects that we can reasonably attempt." Galton called his proposed practices eugenics, from the Greek for "wellborn."

During the late nineteenth and early twentieth centuries, Galton's ideas were taken up not only by racists like Madison Grant but by prominent social reformers like Gifford Pinchot and reproductive rights pioneer Margaret Sanger. The policies supported by eugenicists were as disparate as their motivations—some limited themselves to persuasive measures (prizes for "fitter families") while others supported draconian policies (forced sterilization of people with intellectual disabilities or criminal records). They were united, however, by an outsized belief in the influence of heredity, and in their own ability to choose which families deserved to reproduce. While Spencer had used "fittest" to mean best suited to a situation, not necessarily superior, this subtlety was lost on those for whom eugenics was a state-of-the-art excuse for snobbery. "I wish very much that the wrong people could be prevented entirely from breeding," Theodore Roosevelt lamented in 1914.

As an evolutionary biologist, Huxley knew that destiny was not determined by heredity alone, and he deplored the racist goals of some eugenicists. After the Third Reich took the concept of "race betterment" to its hideous logical extreme—"National Socialism is nothing but applied biology," Hitler's deputy Rudolf Hess declared in 1934— Huxley contributed to a pair of influential United Nations statements that debunked the pseudoscience of racial superiority.

At the same time, Huxley was no egalitarian. He believed that humans had evolved to the point where they could control the evolution of their species, and deliberately elevate its standing in what T. H. Huxley had called "the universe of things." This philosophy of "evolutionary humanism" was, for Julian, a kind of secular religion—a replacement for the faith Darwin had shaken—and it led him to embrace both humanitarian goals and authoritarian means. As a leading member of the British Eugenics Society, he applauded the laws adopted by more than thirty U.S. states that would lead to the "eugenic" sterilization of an estimated sixty thousand people, many of them disabled or mentally ill. He called for the creation of a "single equalized environment" that included improved nutrition and education for the poor. He was an early advocate of expanded access to birth control, but worried that if it were optional, "unteachables" would not use it.

He believed that society should be informed by science and managed by scientists, forgetting that scientists were as capable as anyone else of abusing unbridled power; T. H. might have reminded his grandson that a society run by experts depends on the doubtful ability of "mere human beings" to select the right experts. In 1934, while Hess was using science to justify National Socialism, Julian described his ideal regime in a not entirely tongue-in-cheek book called *If I Were Dictator*.

Aldous Huxley, who was close to his brother but shared none of his affection for centralized authority, described one of Julian's U.S. visits to a friend: "He whirls indefatigably about the country lecturing and having talks with innumerable people about the blue prints of a future society, whose adumbrations, in his essays and letters, fill me, I must confess, with a good deal of gloom."

While Julian Huxley schemed to improve humanity, conservationists were making incremental progress toward Paul Sarasin's vision of *Weltnaturschutz*. Between the world wars, Audubon association president Gilbert Pearson—later the archenemy of Rosalie Edge—worked with European colleagues to create the International Council for Bird Preservation, still operating today as BirdLife International. Conservationists' calls for limits on the freewheeling whaling industry, which was killing tens of thousands of whales each year, helped bring about the first international whaling convention.

In the 1930s Gifford Pinchot, who had been pestering U.S. presidents about the importance of international conservation for decades, found an ally in President Franklin Roosevelt, Theodore's distant cousin. "I repeat again that I am more and more convinced that conservation is a basis of permanent peace," the second Roosevelt wrote to Cordell Hull, his secretary of state. During the Second World War, Roosevelt promised Pinchot that he would convene an international conservation conference once peace was achieved, but he died in 1945—five months before the end of the war, and a year and a half before Pinchot passed away at the age of eighty-one.

In the wake of the war's cascading disasters, many saw the world as much smaller and more interconnected. The relatively obscure word "global" came into wider use, and nations that had once competed for colonial territory began a race to deploy new technology and expand their markets worldwide. The effects of each nation on all the others were clearer than ever before, and for conservationists, alert to ecological relationships, the implications of global economic growth were both worrying and galvanizing. While the eugenics movement fell into disrepute after the Second World War, its Malthusian worldview survived within the conservation movement, and Huxley would encounter it again.

In 1948, Fairfield Osborn—the son of Henry Fairfield Osborn, who co-founded the New York Zoological Society with Madison Grant—published *Our Plundered Planet*, an impassioned warning of the dangers posed by human population growth to soil, water, and species. The younger Osborn, who like the elder was a leader of the Zoological Society, rejected his father's prejudices. "The antipathies of nations and races, the cults of 'superior' and 'inferior' races, cannot be founded on biology," he wrote. Yet even as Osborn substituted planetary peril for racial peril, he echoed his father's dread of the "teeming millions"—the distant but approaching mob, ruled by its reproductive urges.

Another influential book published the same year, *The Road to Survival* by William Vogt, went further. Vogt, who grew up in New York on then-rural Long Island, developed a love of outdoor stories while temporarily paralyzed by polio as a teenager. After studying French literature in college, he worked as a freelance drama critic in New York City, where he pursued his growing interest in birds, falling in with many of the same ornithologists who mentored—and clashed with—Rosalie Edge. He became urgently concerned about other species; hired to edit the Audubon association's magazine, he replaced its odes to nature with reports of environmental ruin, kicking off an internal controversy that ended in his dismissal. Robert Cushman Murphy, the explorer and ornithologist who defended Edge's right to speak at her first Audubon meeting, offered Vogt some consolation in the form of a job studying seabirds in Peru. When he left for South America in 1939, he packed a copy of *Game Management*, the textbook published several years earlier by his friend and mentor Aldo Leopold.

Vogt had been hired by a Peruvian corporation to study bird populations on the guano islands, whose thick strata of bird droppings were lucratively mined for fertilizer. The rocky offshore islands were barren, exposed, and brutally odiferous, but Vogt took to their stark beauty and plunged into his research. Three years of physically punishing observations revealed that shifts in ocean currents caused temporary drops in the supply of plankton and anchovies, forcing millions of adult birds to leave the islands in search of food. "The guano bird population has been withdrawing like a window shade," Vogt wrote to Leopold in the summer of 1939. The consequences were graphic, and for Vogt, unforgettable: hundreds of thousands of seabird chicks, abandoned by their starving parents, dwindled into "pitiful, collapsed, downy clumps."

In the mid-1940s, when Vogt traveled through Latin America as head of the Conservation Section of the Pan American Union, he was appalled by the widespread degradation of forests and farmland. He became convinced that this "land sickness"—a phrase from his ongoing conversations with Leopold—would only worsen as human numbers increased. Once people depleted their resources, he predicted, they would starve like so many clumps of seabird chicks, and the disaster would not be contained. "An eroding hillside in Mexico or Yugoslavia affects the living standard and probability of survival of the American people," he warned in *The Road to Survival*. "We form an earth-company, and the lot of the Indiana farmer can no longer be isolated from that of the Bantu."

For Vogt, the victims of this looming crisis were also the villains. He blamed the "spawning millions" of India and China, and the Puerto Ricans who, he said, "reproduce recklessly and irresponsibly." While he emphasized, with italics, that birth control and sterilization should be *voluntary*—Vogt would later become the director of Planned

Parenthood—he took a Spencerian view of efforts to cure disease and improve sanitation in poorer countries, arguing that such measures were merely "responsible for more millions living more years in increasing misery." Famine, he said, was "not only desirable but indispensable."

Leopold died before the publication of *The Road to Survival,* but in his final years, he too expressed concern about the consequences of human numbers. He was especially interested in the ecological concept of carrying capacity, which he had encountered during his Forest Service days and explored in *Game Management.* Carrying capacity, a term borrowed from shipping, is usually defined by ecologists as the maximum population a given location can support. For Leopold, it was a practical measure of a location's ability to support a species at a particular time. For Vogt, it was a more theoretical concept—shorthand for the maximum number of humans the world could feed, independent of variations in time and place.

Despite these differences, both men saw the collapse of the deer population on the Kaibab Plateau as a grim lesson in carrying capacity. Vogt called it "a phenomenon the human animal should ponder well," and Leopold reminded his readers of "the starved bones of the hoped-for deer herd, dead of its own too-much." Both were convinced that modern humans, preoccupied with short-term economic gain, were testing the carrying capacity of their habitat. For Vogt, the solution was famine. For Leopold, it was the land ethic.

Human population growth was also a central worry for Julian Huxley—and for Aldous, whose *Brave New World* describes a society that, faced with famine, controls its numbers by abolishing pregnancy and parenthood and mass-producing humans through cloning. (The characters in *Brave New World* carry their birth control in Malthusian belts, and dance to the Malthusian blues.) Though Julian corresponded with both Vogt and Osborn, and agreed with many of their views, he optimistically believed that science—and scientific authority—could protect humans from their own too-much.

In a 1965 essay for *Playboy*, Julian recommended that the United Nations "boldly enunciate a global population policy," that every "important" country and region establish "some high-level institution for studying and advising on population problems," and that family planning be incorporated into government services. Through such organized efforts in birth control, he wrote, society could avoid "death decontrol"—"reversion to nature's crude methods of balancing birth and death through famine, disease, and killing."

As Europe began its long recovery from the Second World War, Huxley continued his indefatigable whirling, and soon became involved in discussions about an educational and cultural organization within the newly formed United Nations. He was part of a small group that pushed to include science in the organization's remit, turning its titular acronym of UNECO into UNESCO. In the spring of 1945, the head of the British government's Education Office rather casually asked if Huxley would like to become the acting secretary of UNESCO. Huxley was astonished, and doubtful of his qualifications, but over an extended dinner, he accepted the job.

Huxley had no particular talent for administration—or diplomacy. When he published a pamphlet declaring that his personal philosophy of evolutionary humanism was the guiding purpose of UNESCO, the organization scrambled to distance itself from what the press and public condemned as a form of atheism. But Huxley's broad knowledge and interests, and his enthusiasm for any number of projects, earned him internal support, and in 1946, after a year as acting secretary, he was elected director-general of the organization. He then set about trying to convince its member nations that conservation was part of its mission. At the organization's next general conference, as he grumpily remembered, "Delegates asked what seemed to me silly questions: Why should UNESCO try to protect rhinoceroses or rare flowers? Was not the safeguarding of grand, unspoilt scenery outside its purview? etc., etc." But with "the aid of a few nature-lovers," Huxley recalled, he persuaded the delegates that "the enjoyment of nature was part of culture, and that the preservation of rare and interesting animals and plants was a scientific duty."

In the fall of 1948, Huxley and his allies within UNESCO founded the International Union for the Protection of Nature (later the International Union for the Conservation of Nature), a network of governments, government agencies, and nongovernmental organizations that was assigned to "collect, analyse, interpret, and disseminate information" about conservation. At the organization's founding conference, held at the palace of Fontainebleau on the outskirts of Paris, Huxley paid poetic tribute to this somewhat bureaucratic mission, speaking of "the fascination of all these other manifestations of life which, though all products of the same process of evolution, yet are something in their own rights, are alien from us, give us new ideas of possibilities of life, can never be replaced if lost."

Since childhood, Huxley had delighted in familiar and exotic forms of life, finding wonder in them both. The IUCN, his brainchild, was the first intergovernmental association dedicated to their survival.

Until this point, conservationists had devoted most of their energy to keeping common species common—either because they liked to hunt them or because they saw that they were being cruelly exploited. With the prominent exception of the bison, the extinction of treasured species was a motivating but relatively distant threat. The IUCN would show the world that many species were closer to extinction than anyone realized—and would, fatefully, shift the focus of the conservation movement to these emergency cases.

The newborn IUCN faced what historian Robert Boardman calls "a data gap of awesome proportions." As Leopold had observed twelve years earlier in his essay "Threatened Species," existing information about vulnerable species of all kinds was incomplete and "scattered in many minds and documents." By the time the union tasked itself with the not-so-modest goal of conserving "the entire world biotic community," the situation was not much better.

Resolving that they had to start somewhere, IUCN advisers drew on existing studies by the New York Zoological Society to compile a list of fourteen mammals and thirteen birds thought to be most in danger of extinction. (Eight of those twenty-seven species and subspecies are now believed to be extinct; one has been "lumped" with another by taxonomists; almost all the rest, including the whooping crane and the California condor, are still highly vulnerable, though in most cases their numbers have been boosted by captive breeding and reintroduction efforts.) In 1954, the union hired a young ecologist from the United States, Lee Talbot, to carry out an international version of Leopold's game survey. Talbot, the grandson of zoologist C. Hart Merriam and grand-nephew of pioneering birdwatcher Florence Merriam Bailey, spent six months traveling through the Middle East, southern Asia, and northern Africa as a surveyor-cum-ambassador, talking to anyone with anything to say about the state of local wildlife. His report, *A Look at Threatened Species*, was an energetic travelogue that not only established a scientific baseline for the organization but drew public attention to its mission.

In 1963, Peter Scott—the son of the ill-fated Antarctic explorer— became head of the IUCN Species Survival Commission. (Scott, an ornithologist and artist, had almost single-handedly saved one of the bird species on the initial IUCN list, the Hawaiian goose, or nēnē, by breeding the birds in England and shipping thirty-five geese to Hawaii for release.) Under Scott's direction, the commission compiled work by Talbot and others into a series of profiles of threatened species. What began as a stack of index cards turned into a binder of loose-leaf pages describing 277 mammals, distributed to specialists in 1966 as the first of the soon-to-be-iconic Red Data Books.

Today, there are more than one hundred thousand species of animals and plants on what is known as the IUCN Red List, and scientists classify and periodically reclassify its species into categories ranging from Least Concern to Extinct. The list makes no distinction between "useful" species and others; Rosalie Edge and Aldo Leopold would be pleased to see hawks, wolves, and beetles alongside whales and bison.

It underlies international agreements such as CITES, the Convention on International Trade in Endangered Species of Wild Fauna and Flora, which went into effect in 1975. It is also an influential global barometer, used by conservation organizations large and small to set priorities and measure progress and reversals.

Huxley's determined push for a conservation program within UNESCO was successful, but he had earned some lasting enemies within the organization, and after two years as director-general the board voted not to extend his term. ("On the beautiful shores of the Bosphorus / Our Board's being rather prepospherous," he wrote dejectedly during a meeting in Turkey near the end of his tenure.) He continued his work with UNESCO, however, and in 1960, the organization sent him on a three-month trip through ten central and east African countries. Juliette accompanied him, and he described their experiences in a vivid three-part series published in the London *Observer*:

> And the sight of great herds of topi or gnu or zebra galloping across the open plains, of a troop of elephants coming down to drink and play, of a pride of lions on a kill, of sausage-like hippos in and out of the water, of a herd of impala leaping in all directions, of pre-history incarnate in a rhinoceros, of a family of giraffes cantering along like elongated rocking horses—any of these is unforgettable, a unique contribution to the riches of our experience.

Like many European and North American conservationists at the time, Huxley was worried about the survival of these and other African species, and the immediate source of his concern was the independence movement gathering momentum across the continent. Conservation in Africa had so far been a colonial project, and not a few European conservationists feared, as Huxley put it in the *Observer*, "that with the inevitable advent of African governments in most of the territories, game will be regarded as so much meat conveniently provided on the hoof . . . that National Parks will be looked upon as unwanted relics of 'colonialism' or as a silly European invention, of no value to the up-and-coming African States, and that no large wild animals will be allowed to survive in African nature."

Huxley was more confident than most, however. While he patronizingly believed that newly independent Africans would need to be taught the value of their continent's wildlife—as a source of protein, tourist dollars, national pride, and international respect—he trusted that African leaders, and African people, would quickly realize and embrace the benefits of conservation.

Huxley's *Observer* series caught the attention of many, including Victor Stolan, a Czech businessman who had come to Britain as a refugee. He wrote earnestly to Huxley, "There must be a way to the conscience and the heart and pride and vanity of the very rich people to persuade them to sink their hands deeply into their pockets and thus serve a cause which is greater and nobler than any other one—absolutely." Huxley agreed, and he passed the letter on to IUCN co-founder Max Nicholson. Inspired by Stolan, Nicholson proposed to create a private fundraising arm for the IUCN, which was struggling to cover the costs of its work.

Over several months in 1961, Nicholson, Huxley, and some two dozen other naturalists—mostly British and all male, with the exception of ornithologist Phyllis Barclay-Smith—held a series of planning meetings in London. By the third meeting, the group had agreed to name the organization the World Wild Life Fund. By the sixth meeting, it had decided on its mascot: a cuddly, threatened animal who would emphasize the global reach of the fund. Conveniently, the species also had black-and-white fur, making it cheap to portray in print, and it took Peter Scott only a few minutes to draw the group's famous panda logo. The World Wildlife Fund, which has since become an independent organization, now works in more than one hundred countries.

The pessimists whose views Huxley described in the *Observer* were at least partly right; many of the people who lived on the vast and varied African continent did see national parks as relics of colonialism, and conservation as a European invention. How could they not? The first national park in Africa, now called Virunga National Park, is said to have gotten its start around a Yellowstone National Park campfire in 1919, when the elder Henry Fairfield Osborn helped convince King Albert of Belgium that he should protect mountain gorilla habitat in what is now the Democratic Republic of the Congo. Most of the rest of the continent's parks and game reserves had likewise been created by colonial governments, many of which proceeded to forcibly evict "squatters" from places where they had lived for centuries or more— much as the creators of Yellowstone and Yosemite had evicted Native Americans from the newly designated parks.

Long before colonization, many African societies had established conservation practices ranging from royal edicts to informal customs. Like human societies throughout the world, they valued other species in multiple ways. But many Africans came to believe, for good reason, that the conservation measures supported by colonial governments and international groups were intended to reserve the continent's species for foreigners—and prevent Africans from using the resources they regarded as their birthright. "The question African people ask is: for whom are we conserving wildlife?" Zambian zoologist Malumo Simbotwe wrote in the early 1990s. For much of the previous century, the answer had been all too obvious.

Early encounters between the conservation movement and newly independent African nations did little to allay these concerns. The World Wildlife Fund recruited Prince Philip as president of its United Kingdom branch, and Prince Bernhard of the Netherlands as the ceremonial president of the organization—glittering associations designed to impress potential donors in Europe and North America, not reassure potential partners in Africa. When the WWF made its international debut at an IUCN conference in Tanganyika—part of present-day Tanzania—in 1961, Nicholson and Huxley asked Julius Nyerere, the popular young prime minister of the soon-to-be-independent country, to read a "Declaration of a State of Emergency" that they had composed. Though Nyerere had previously expressed support for conservation goals, he balked at the implication that independence constituted an "emergency" for African wildlife. Nyerere eventually agreed to endorse a milder statement, written by Nicholson, saying that the "conservation of wild life and wild places not only affects the Continent of Africa but the rest of the world as well."

Nyerere and other leaders initially welcomed WWF and IUCN efforts to encourage tourism and provide technical assistance to national parks and game departments. But many foreign conservationists saw the African landscape as John Muir had seen Yosemite—as an extraordinary place meant to be visited, not lived in. The fact that much of Africa was already being managed for human needs, in ways Gifford Pinchot might have recognized, meant little.

German veterinarian and conservationist Bernhard Grzimek, the director of an influential 1959 documentary called *Serengeti Shall Not Die*, declared that "not even natives" should be permitted to live in a "primordial wilderness" like the Serengeti. Representatives of the WWF and IUCN in East Africa pushed for the exclusion of nomadic Maasai herders from national parks and game reserves, shutting them out of their traditional hunting and grazing territory. By the end of the 1960s, tensions between conservationists and the Maasai were so high that some Maasai reacted to a new national park proposal by slaughtering rhinos in protest.

Huxley had helped to expand the reach of modern conservation, creating international institutions that could work on broad ecological scales. His work made it possible for conservationists to not only know and care about species in faraway places, but take concerted steps to protect them. Because of his faith in centralized power and the wisdom of experts, however, he saw conservation primarily as a top-down enterprise, with regional and local institutions executing the orders of a global authority. If the conservation movement is in a sense its own biotic pyramid, in which change cycles from top to bottom and bottom to top, Leopold emphasized the bottom layer; Huxley emphasized the top.

Like many conservationists of his time and place, Huxley was especially dismissive of local and regional institutions in Africa—which had been weakened by centuries of colonization, and would be further eroded by the autocrats and dictators who gained power in the decades after independence. (Nyerere, a committed socialist, was admired at home and abroad for his moral leadership but criticized for his embrace of one-party rule, under which he served as president for more than twenty years.) Not until the 1980s would a new generation of African conservationists recognize what had been lost, and set about restoring the foundation of the conservation pyramid.

Huxley would remain active in public life until shortly before his death in 1975, at the age of eighty-seven, and in the fall of 1963 he traveled to Nairobi for the eighth General Assembly of the IUCN. Scheduled to address the delegates was Stewart Udall, the U.S. Secretary of the Interior and one of the youngest members of President John Kennedy's youthful Cabinet. Just three days before his speech, Udall was struggling toward the nineteen-thousand-foot peak of Mount Kilimanjaro. An experienced outdoorsman, Udall had insisted on including the climb in his already crowded schedule, and to save time he was attempting to reach the summit in three days instead of the recommended four.

During the ascent, one of Udall's guides, a young Tanganyikan army captain named Mrisho Sarakikya, collapsed and had to be given oxygen by a rescue team. Udall himself barely staggered to the top. "Looking back, it was one of my most pleasant trips—except possibly for the final 1,500 feet," *Life* magazine photographer Terry Spencer dryly recalled to Udall. At the summit, the climbers were crowned, according to tradition, with garlands of tiny everlastings; Udall and Spencer, flowers in their hair, were photographed embracing and grinning at one another, giddy with relief. (Sarakikya would become the founding commander of his country's civilian-controlled defense forces, and would go on to climb Kilimanjaro more than forty times.)

After Udall descended, he squeezed in an aerial tour of the Rift Valley before arriving in Nairobi for his address. Perhaps chastened by his near-disaster on Kilimanjaro, he struck a humble note. The United States, he said, was the country that had nearly exterminated the bison, created the Dust Bowl, and extinguished the passenger pigeon. It was still trying to repair the damage it had done to the Florida Key deer, the manatee, the Eskimo curlew, and the whooping crane. "I have unveiled

Stewart Udall (right) and photographer Terence Spencer (left) celebrating on the summit of Mount Kilimanjaro, September 1963.

these sorry episodes in our conservation history in order to make a plea today to the emerging nations to study our mistakes as well as our achievements," he said.

The new African nations had an opportunity, Udall said, to "preserve their treasures of natural beauty and habitat . . . while simultaneously exploiting those natural resources that can produce material abundance." Udall added that the continent's justly famous parks and reserves, the success of game-ranching enterprises in Southern Rhodesia, and the new, internationally funded College of African Wildlife Management in Tanganyika were all encouraging signs that African nations could avoid his own country's pitfalls and follow "more fruitful paths to the wise use of nature's bounty." Huxley, in the audience, must have nodded in agreement.

Udall had a longtime interest in conservation, and a recent bestseller called *Silent Spring*—and his conversations with its author, Rachel Carson—had deepened his commitment to protecting nature's bounty. He left Africa depleted from his climb, but further inspired. Back in Washington, DC, he began to lay the foundations of the most powerful species-protection law in the world.

Chapter Six

THE EAGLE AND THE WHOOPING CRANE

"They came by like brown leaves drifting on the wind," a visitor to Hawk Mountain wrote in the fall of 1945. "Sometimes a lone bird rode the air currents; sometimes several at a time, sweeping upward until they were only specks against the clouds or dropping down again toward the valley floor below us; sometimes a great burst of them milling and tossing, like the flurry of leaves when a sudden gust of wind shakes loose a new batch from the forest trees."

The visitor was Rachel Carson, then thirty-eight and serving, somewhat reluctantly, as a writer and editor for the U.S. Fish and Wildlife Service. Carson dreamed of supporting herself as a full-time author, but her first book, *Under the Sea-Wind*, had sold poorly, and she couldn't afford to leave the agency. She consoled herself with occasional magazine work, and after her visit to Hawk Mountain, the sanctuary founded by Rosalie Edge, she began to write an article about it.

Carson's background was in marine science, and her affections were with the sea. Hawk Mountain was not her home landscape, but during that cold, blustery morning on its sweeping ridgeline, she heard the call of the ocean. "Perhaps it is not strange that I, who greatly love the sea, should find much in the mountains to remind me of it," she wrote. "I cannot watch the headlong descent of the hill streams without remem-

Rachel Carson watching migrating hawks at Hawk Mountain, October 1945.

bering that, though their journey be long, its end is in the sea . . . Now I lie back with half closed eyes and try to realize that I am at the bottom of another ocean—an ocean of air on which the hawks are sailing." A photo from that morning, taken by her friend and fellow bureaucrat

Shirley Briggs, shows Carson perched on a boulder at the sanctuary's North Lookout, dressed like a storybook aviatrix in a leather jacket and wide trousers, binoculars raised to the sky.

Carson never published her article on Hawk Mountain, and her notes would remain buried in her papers for decades. But she remembered her visit and, fifteen years later, from her home in Silver Spring, Maryland, she sent a letter to sanctuary caretaker Maurice Broun. "As you may possibly have heard from some of our friends in the conservation world, I am at work on a book that will explore some of the effects of chemical pesticides, especially their ecological effects," she wrote. She continued:

> In writing of what has happened to the birds, I am of course mentioning the problem of the eagles even though we cannot as yet pin it down specifically to pesticides. I have seen you quoted at various times to the effect that you now see very few immature eagles in fall migration over Hawk Mountain. Would you be good enough to write me your comments on this, with any details and figures you think significant? . . . You may be sure that anything you tell me will be most useful.

Broun's answer would become a key piece of evidence in *Silent Spring*, Carson's legendary argument against the overuse of synthetic pesticides. And *Silent Spring*—along with Carson herself—would illuminate a new kind of threat to the world's species, and catalyze a new movement to protect them.

Even before her first visit to Hawk Mountain, Carson had been thinking about the effects of pesticides on other species. She was especially interested in dichloro-diphenyl-trichloroethane, or DDT, which had been synthesized in the 1870s but remained relatively unknown until the late 1930s, when Swiss chemist Paul Müller discovered that it was a remarkably effective insect-killer. Early studies suggested that it had little effect on warm-blooded animals, including humans, and during the Second World War, when Allied countries lost access to the Japanese-grown chrysanthemums that had been a primary source of the insecticide pyrethrum, U.S. and British forces enthusiastically adopted DDT. "We have discovered many preventives against tropical diseases and often against insects of all kinds," British prime minister Winston Churchill announced in September 1944. "The excellent DDT powder, which has been fully experimented with and found to yield astonishing results, will henceforward be used on a great scale."

Allied soldiers sprinkled themselves with DDT to ward off lice and mosquitoes, and fitted torpedo bombers with spray nozzles in order to "strafe" Pacific islands with a solution of the pesticide in diesel oil. In Naples, Italy, in 1943, U.S. health officials directed the treatment of more than a million residents with DDT in order to contain an outbreak of typhus. "Neapolitans are now throwing DDT at brides instead of rice," reported the *New York Times*. (Photos from the time, which show U.S. soldiers spraying DDT powder at raggedly dressed residents, suggest a less celebratory atmosphere.)

On the U.S. home front, DDT gained near-mythical status as a symbol of national power and ingenuity, and was promoted as a patriotic weapon in the war against disease and household pests. One advertisement showed Uncle Sam vanquishing Japanese prime minister Hideki Tōjō with one hand and mosquitoes with the other. DDT was sprayed on forests, farm fields, and neighborhoods to fight maladies ranging from gypsy moths to malaria; one Illinois community doused itself with DDT in a misguided attempt to control polio. Brigadier General James Simmons, a senior army surgeon, wrote in the *Saturday Evening Post* that DDT was "the War's greatest contribution to the future health of the world." In 1948, Müller was awarded the Nobel Prize in Medicine.

Carson knew that scientists were discovering another side of DDT. In June 1945, researchers with the Fish and Wildlife Service used a small plane to spray a mixture of DDT, xylene, and fuel oil on a tract of forest in the Patuxent Research Refuge in Maryland, then monitored the mammals, birds, frogs, and fish in the forest and the nearby Patuxent River. Within ten hours, dead fish began to collect in a net stretched across the river. Later laboratory studies by agency scientists showed that when birds, amphibians, and small mammals were directly exposed to DDT, most sickened and many died—all trembling, twitching, and eventually collapsing from what appeared to be catastrophic damage to their nervous systems. While DDT in powder form was not easily absorbed through the skin, it appeared that DDT dissolved in oil was, and that absorbing or ingesting the pesticide was fatal to many species.

Carson wrote some of the press releases that described these results, and she proposed a story to *Reader's Digest*, circumspectly describing the experiments as having "more than ordinary interest and importance." The magazine turned down the idea, but Carson continued to follow the science.

Over the next decade, the anecdotes—and the data—piled up. Aerial spraying of DDT over the Miramichi River in New Brunswick, Canada, killed more than 90 percent of the river's young salmon. Spraying over the Yellowstone River in Montana decimated the river's young trout. After forests and small towns were sprayed with DDT in an effort to control Dutch elm disease, reports of dead and dying birds poured in. In 1948, when a group of golf courses north of New York City treated their greens with a concentrated solution of DDT, Rosalie Edge reported to the director of New York's state fish and game department that "nineteen robins were found dead on a small lawn. At another place three orioles were found dead, in spite of the arboreal habits of the oriole. I might go on giving you many examples."

Researchers discovered that both DDT and one of its breakdown products, DDE, could scale the biotic pyramid. A study in Michigan suggested that as elm leaves treated with DDT fell to the ground, DDT and DDE moved from the leaves into the soil, from the soil into the bodies of earthworms, and from the earthworms into the bodies of robins, where the toxins accumulated in lethal concentrations. Scientists suspected that they were also accumulating in humans.

In the fall of 1957, Carson's attention turned to a legal dispute over DDT on Long Island, where a group of residents—led by ornithologist Robert Cushman Murphy—had sued state and federal agriculture officials over an aerial-spraying campaign against gypsy moths. The plaintiffs claimed that the spraying of DDT in their neighborhood endangered health and property, for the pesticide was "recognized and admitted by the defendants to be a delayed-action, cumulative poison such as will inevitably cause irreparable injury and death to all living things."

After a twenty-two-day trial, the judge ruled against the plaintiffs, concluding that the Long Island residents had "presented no evidence that they or anyone else were made ill by the spraying of DDT." The residents eventually took their appeal to the U.S. Supreme Court, which voted not to review the case. Outside the courtroom, however, their arguments proved persuasive. New York State halted its spraying program, and Carson, who had begun a lively correspondence with plaintiff Marjorie Spock—an organic farmer and the younger sister of famed pediatrician Benjamin Spock—decided that it was time to write a book about DDT.

Carson, a meticulous researcher, pieced together the information she had and plied her contacts for more. The data Maurice Broun sent in response to her 1960 letter proved, as she later wrote, "especially significant." Between 1935 and 1939, the first four years of the daily bird counts at Hawk Mountain, some 40 percent of the bald eagles Broun observed were young birds. Two decades later, however, young birds made up just 20 percent of the total number of bald eagles recorded, and in 1957, Broun counted only one young eagle for every thirty-two adults. Broun's observations were the longest and most detailed record of a more and more familiar pattern: birds were occupying fewer nests, producing fewer eggs, and raising fewer young— partly because, as later research would show, the accumulation of

DDT breakdown products in their bodies caused them to lay eggs with shells too fragile to incubate.

When *Silent Spring* was published as a three-part series in *The New Yorker* in the summer of 1962, the reaction was immediate. Public anxiety about nuclear war—including the insidious effects of nuclear fallout on the human body—was already acute, and Carson drew an explicit parallel between the invisible, existential threats of atomic radiation and those of synthetic pesticides. (An ardent believer in nuclear nonproliferation, Carson dedicated *Silent Spring* to the physician and anti-nuclear activist Albert Schweitzer.) The comparison would continue to resonate; the hardcover edition of the book was published just one month before the Cuban Missile Crisis, and one year before the first nuclear test ban.

Though Carson's detractors belittled the book as an "emotional and inaccurate outburst" and the cause of needless public alarm—"Silence, Miss Carson!" scolded the headline of one review—her case rested on a body of evidence, not on any single study or example, and she was careful to emphasize the uncertainties in the science she cited. Contrary to a lasting misperception, she also readily acknowledged that in some cases chemicals could be useful in protecting crops and limiting insect-borne diseases. "I favor the sparing, selective and intelligent use of chemicals," she said. "It is the indiscriminate, blanket spraying that I oppose." She wrote with clear elegance, and while she avoided overtly political arguments, she delivered her call for moderation with moral force. "Who has decided—who has the *right* to decide—for the countless legions of people who were not consulted that the supreme value is a world without insects, even though it be also a sterile world ungraced by the curving wing of a bird in flight?"

While others had called public attention to the far-reaching effects of pesticides, none had done so as powerfully as Carson. Her brief opening chapter, "A Fable for Tomorrow," dropped readers into a future in which no birds sang, no eggs hatched, no flowers turned to fruit. Children at play were stricken with sudden illnesses, and died within hours.

Like many of Aesop's fables, Carson's was a warning against ignorance, but the consequences she described were both personal and global. The poison killed the swallows, swept away the predatory insects that "play the same role as the wolves and coyotes of the Kaibab," and seeped into human organs—cascading around the world, from one species to the next. "We are losing half the subject-matter of English poetry," Aldous Huxley said to Julian after reading Carson's book.

Carson, like the Huxley brothers, understood that both evolution and ecology often bore unsettling tidings, but she urged people to listen. In her last public address, which she gave to an audience of doctors and other health care providers in San Francisco in early 1963, she noted that Victorians faced with the theory of evolution eventually "freed themselves from the fears and superstitions" that made them recoil from its implications. "Yet so many of us deny the obvious corollary," she said, "that man is affected by the same environmental influences that control the lives of all the many thousands of other species to which he is related by evolutionary ties." Humans could extinguish other species, and be extinguished themselves—and some means of doing so, Carson warned, were invisible and inescapable, nowhere and everywhere at once.

Silent Spring attracted millions of readers, but two of them had more influence than most: U.S. president John Kennedy and his Secretary of the Interior, Stewart Udall. Kennedy, a yachtsman, loved to sail near his family home on Cape Cod, Massachusetts, and he kept Carson's books *The Edge of the Sea* and the bestselling *The Sea Around Us* on his bookshelf. Kennedy's interest in conservation was fitful, to Udall's chagrin, but he did care about the ocean and the seashore, and Carson's work had moved him.

Both Kennedy and Udall were aware of *Silent Spring* before its publication, and in August 1962, after it was serialized in *The New Yorker*, Kennedy acknowledged the strength of Carson's case, stating in a televised press conference that "Miss Carson's book" had prompted the administration to look more closely at the effects of pesticides on human health. The next day, Kennedy announced that a panel of the President's Science Advisory Committee would study the issue; the panel eventually affirmed Carson's findings.

Udall, who came from a politically prominent Mormon family in rural Arizona, had been raised with a sense of obligation to the land, and he was committed to pursuing what he called a "vigorous program" at the Department of the Interior. He saw *Silent Spring* as an opportunity to put pesticide regulation on the department's agenda, and assigned one of his closest advisors, Paul Knight, to coordinate with Carson throughout the book's launch. After publication, when Carson was attacked by both the chemical industry and its sympathizers within the government, Udall himself advised and supported her. "Rachel was a friend," Udall remembered. "She was a courageous woman. She stood up—*by herself.* She didn't come running to me and say, 'You've got to speak up, Udall.'"

In April 1963, at the dedication of a new wildlife research center at the Patuxent refuge, Udall praised Carson as a "great woman" who "awakened the Nation by her forceful account of the dangers around us." Days later, Carson wrote to thank him. "During the years I worked on *Silent Spring*, there were times when I wondered whether the effort was worth while—whether the warnings would be heeded enough to change the situation in any way," she wrote. "I can truthfully say that nothing has pleased me more than the tribute you paid in dedicating the laboratory." Her next letter to him ended less formally: "Please give my greetings to your nice family, including, of course, Dreamcat."

Carson was ambitious, and resolute in the face of public controversy, but she was also acutely sensitive, highly attuned to other species and to people of all ages. Stanley Temple, who in 1976 would step into Aldo Leopold's shoes as a professor of wildlife ecology at the University of Wisconsin at Madison, met the woman he knew as Miss Carson when he started tagging along on Audubon Society field trips around

Washington, DC, as a young boy. "Most of the adult naturalists I knew wanted to teach me to identify things," he told me. "She taught me to stop and look."

Private and inclined to solitude, Carson never married, but for the last dozen years of her life she enjoyed an intensely romantic friendship with Dorothy Freeman, her summer neighbor on the coast of Maine. "Somehow the sharing of beautiful and lovely things is so much more satisfying with *you* than it has ever been for me with anyone else," Carson wrote to Freeman shortly before Freeman's birthday in June 1955. Carson continued:

> May the year ahead be filled with deep happiness for you: with moonlight on the shore and the distant sound of rote, bunch-berries in the spruce woods, sun shining through drifts of apple blossoms, the song of veeries at twilight and hermits at dawn; with laughter over breakfast coffee and sherry by the fire; explorations of fairy pools; freesias and white hyacinths—oh, you know the list—all this and more. And darling, may the year deepen the growing understanding between us—and may I, who love you so dearly, bring you only happiness.

In April 1960, just before she turned fifty-three, Carson underwent a radical mastectomy to remove two tumors from her left breast. Though her surgeon gave her the impression that the operation was only precautionary, she suspected that her situation was more serious, and in December a specialist confirmed that she had cancer. She hid her illness from all but Freeman and a few other confidants, fearing that the chemical industry would say the disease had prejudiced her against DDT. Weakened by radiation treatment and multiplying ailments, she labored to finish *Silent Spring*—and then, after its publication, to withstand the controversy it created. When journalist Eric Sevareid interviewed her on camera for *CBS Reports* in late 1962, Carson wore a heavy wig and looked so ill that Sevareid worried she would not survive to see the story air. (She did, and her impeccable performance served to counter industry criticism.)

Carson died in April 1964, and Udall served as a pallbearer at her funeral. After the memorial service, a friend of Carson's handed him a piece of paper printed with a passage from Carson's work. The friend explained that Carson had wanted the lines to be read at her service, but Carson's brother had said no. Udall, appalled that one of Carson's final wishes had been ignored, put the paper in his pocket.

Udall outlived Carson by nearly fifty years. During his eight-year tenure at the Department of the Interior, he established four national parks, fifty wildlife refuges, and eight national seashores. Dubbed the "Secretary of Things in General" by the *Saturday Evening Post*, he tackled issues ranging from historic preservation to water quality to civil rights, demanding at one point that the Washington, DC, football franchise integrate its team before playing in a stadium located on federal land. He supported both dam projects and wilderness, believing—at least at the time—that conservationists could back both. Shortly before President Lyndon Johnson left office in early 1969, he sent Udall a heartfelt if slightly exasperated letter of thanks: "I have no doubt that you have been one of the most active, colorful, and productive Interior Secretaries in history."

Later in life, Udall returned to the southwestern United States, where he sued the federal government on behalf of Navajo uranium miners sickened by radiation exposure. The ultimate failure of the case shook his faith in government, but he successfully lobbied for congressional investigations that led to compensation for thousands of people affected by radiation poisoning. He also remained active in conservation, and repeatedly expressed his respect for Carson and her ideas.

In 2007, at a centennial celebration of Carson's birthday at the John F. Kennedy Presidential Library in Boston, Udall stood straight-backed at the podium and told the crowd that he was going to read the words Carson chose for her memorial service. At eighty-seven, Udall had lost much of his vision, but with the quiet assistance of his college-aged grandson, he slowly read the passage from Carson's third book, *The Edge of the Sea*:

> Contemplating the teeming life of the shore, we have an uneasy sense of the communication of some universal truth that lies just beyond our grasp. What is the message signaled by the hordes of diatoms, flashing their microscopic lights in the night sea? . . . And what is the meaning of so tiny a being as the transparent wisp of protoplasm that is a sea lace, existing for some reason inscrutable to us—a reason that demands its presence by the trillion amid the rocks and weeds of the shore? The meaning haunts and ever eludes us, and in its very pursuit we approach the ultimate mystery of Life itself.

Until the midpoint of the twentieth century, the word "environment" was usually a synonym for "context." The modern concept of a planetary life-support system called "the environment" grew out of the sense of global connection—and global vulnerability—that so many felt after the Second World War. *Silent Spring* gave shape and direction to these anxieties, and by doing so helped launch the environmental movement.

The terms "environmentalism" and "conservation" are often used interchangeably, and the movements are largely complementary; both strive to protect the rest of life from human excesses. But their pedigrees differ, and so do their priorities. Early environmentalists brought middle-class attention and international perspective to problems like air and water pollution, long left to labor activists and urban reformers. Conservation, meanwhile, was shaped by its beginnings in elite hunting circles, and despite the broadening influences of agitators like Rosalie Edge, ecologically minded scientists like Aldo Leopold, and internationalists like Julian Huxley, it continued to emphasize the protection of charismatic species from immediate and visible threats. For conservationists of Carson's time, protecting other species was often an end in itself; for environmentalists, it was part of a larger mission.

Conservation was also founded on the belief that sport hunting was a worthwhile pursuit—a belief lost on many environmentalists, including Carson herself. When Oxford University Press requested that Carson write a promotional quote for *Round River*, a collection of Leopold's hunting tales, she responded with a blistering letter. "Mr. Leopold was a completely brutal man," she wrote. "Oxford has done a service in revealing one of the things that is wrong with conservation—that so much of it is in the hands of men who smugly assume that the end of conservation is to provide fodder for their guns—and that anyone who feels otherwise is a sentimental fool." Carson noted, in the margin of the letter, that she had written her rebuke without reading the rest of Leopold's work ("for *Sand County Almanac* came when I was too busy with *Sea* to do more than skim it"). Disagreements between hunters and those Leopold had called "protectionists" would persist, however, to the detriment of their common interests.

Udall understood that the environmental fable Carson told in *Silent Spring* was about more than the singular problem of DDT and its threat to individual species, and he took its challenge seriously. His book *The Quiet Crisis* called for a "land ethic for tomorrow" that was "as comprehensive as the sensitive science of ecology." At the Department of the Interior, he aimed to improve what he called the "total American environment," beautifying city centers as well as establishing national parks. He argued that the national character owed its best qualities to an ongoing relationship with the land, not—as Frederick Jackson Turner's frontier thesis implied—to domination of the land. Udall also called for an end to the old conflict between Muir-inspired preservationists and Pinchot-inspired utilitarians, exhorting those on both sides to aspire to a "higher order of conservation statesmanship."

When Udall traveled to East Africa, in the fall of 1963, Carson's ideas were on his mind. "Environment preservation . . . is infant science that must engross the talent and energy of best minds," he scrawled on a typewritten draft of his speech to the IUCN conference in Nairobi. "Widening concept of conservation to cover *all* resources and *all* resource problems!" On September 15, the day after Udall descended Mount Kilimanjaro, the *New York Times Magazine* published his essay "To Save Wildlife and Aid Us, Too," in which he named what he saw as the three most pressing human threats to other species: overhunting, habitat destruction, and air and water pollution. "The entry of numerous poisons into the chain of life may be a fateful event for wildlife species—and perhaps for some members of the human species as well," he wrote, nodding to *Silent Spring*.

It was high time, he argued, to protect wildlife not only from direct threats such as unregulated hunting but from "the side effects of advancing civilization," including the obliteration of open space and "the widening circle of pollution." His attempts to widen the concept of conservation, however, would be hindered by the history of the movement—and by a desperate bid to save a species that, as Udall wrote in the opening paragraphs of his essay, "symbolizes the plight of vanishing types of birds and animals everywhere": the whooping crane.

"We have cranes of two colors," the Jesuit priest Pierre François Xavier de Charlevoix wrote during his travels through the French territories of North America in the early 1700s. "Some are all white, and the others are grayish. They all make excellent soup."

Charlevoix's senses may have been sharpened by his appetite, for his account is one of the few colonial-era reports that clearly distinguishes between the two species of North American cranes: the pale gray sandhill crane, *Antigone canadensis*, and the white, far less common whooping crane, *Grus americana*. Sandhills are about three-quarters the size of whoopers, which are the tallest birds in North America, but the species occasionally travel together, and they have a similarly striking silhouette. Philip Amadas, an English sea captain who landed on the coast of present-day North Carolina in 1584, might or might not have been describing whooping cranes when he reported that after his men fired their muskets, a flock of "the most part white" cranes "arose, with such a crye, re-doubled by many echoes, as if an armie of men had show-

ted altogether." The eminent English botanist and zoologist Thomas
Nuttall, who described "numerous legions" of whooping cranes flying
above him as he descended the Mississippi River in 1811, was likely
misled by what a later biologist politely called an "enthusiastic imagination." Even John James Audubon, who included a vivid and unmistakable portrait of a whooping crane in *The Birds of America*, mixed up the
two crane species during his travels in the early 1800s.

Unlike the bison and the passenger pigeon, the whooping crane was
probably never a terribly abundant species. Fossils are scarce; scientists
now estimate that there were some ten thousand whooping cranes in
North America before European settlement, and that by the mid-1800s
hunting and habitat loss had reduced the population to fewer than fifteen hundred birds. In 1858, the species was dealt another blow when
ornithologist Spencer Fullerton Baird, then the assistant secretary of
the Smithsonian, warned that the whooping crane, "though common
in Texas and Florida, is yet one of the rarest birds in collections." It
was as if an armie of men had showted altogether: amateur and professional collectors alike loaded their weapons and went looking for
whooping cranes, eager to acquire skins and eggs for their natural history cabinets. By 1913, there were fewer than one hundred whooping
cranes left on the continent, and William Hornaday predicted that "this
splendid bird will almost certainly be the next North American species
to be totally exterminated." In 1929, with numbers still dropping, Willard Van Name declared that the species was "beyond saving."

The whooping crane crisis attracted public attention. The Carolina
parakeet, whose jewel-like plumage so tempted ornithologist Frank
Chapman, was nearly if not entirely extinct. On the island of Martha's
Vineyard, off Cape Cod, the world's only known population of heath
hens had been reduced to a single bird. Legions of North American
bird lovers had recently fought to restrict the plume trade and regulate
bird hunting; that the stately, ancient whooping crane was nonetheless
about to follow the parakeet and the passenger pigeon into extinction
was, to many, an outrage.

Neither sandhill cranes nor whooping cranes had fared well in nineteenth-century North America. Both large, slow-moving species were easy targets for commercial and sport hunters, and their flesh was considered so delicious that they were nicknamed "the ribeye of the sky." Both were disadvantaged by the conversion of wetlands to farmland. But after bird hunting was restricted in the early twentieth century, sandhill populations began to recover, not only near the Leopold shack in Wisconsin but across the continent. Whoopers—which were less abundant to begin with, and are less flexible in their diet and habitat—continued to suffer.

In 1937, when the whooping crane count hovered around thirty birds, President Franklin Roosevelt established what is now the Aransas National Wildlife Refuge, a 180-square-mile chunk of salt marshes and sandy prairie on the Texas coast that was the wintering ground of more than half the remaining whoopers. The move came just in time; within five years, the migratory population was down to fifteen birds, and all of them wintered at Aransas.

Each spring, the birds migrated north, and each fall they returned to Aransas, usually with a handful of chicks in tow. But no one knew exactly where the last remaining whooping cranes spent the summer, and every year, refuge managers feared the birds would not return at all. In 1945, the newly created U.S. Fish and Wildlife Service partnered with the National Audubon Society on the Whooping Crane Project, an effort to better understand the species and protect its future.

At the time, Audubon ornithologist Robert Porter Allen was living in remote Tavernier, Florida, doing battle with sandflies as he clambered through the mangrove forests of the Florida Keys. Allen had loved birds since his childhood in rural Pennsylvania, but he preferred action to academics, and after two restless spells at college he left to join the merchant marine. Following four years at sea, he landed in New York City, married Evelyn Sedgwick—a classically trained pianist with a sense of adventure—and talked himself into a job at the headquarters of the National Association of Audubon Societies. He quickly graduated from menial tasks to fieldwork, and the society posted Allen and his family to the Keys. There, Allen began to study the habits of roseate spoonbills—skittish birds famous for their flame-colored plumage, but nearly unknown otherwise.

Ecologists had recently started to use the term "ecological niche" to describe the food, habitat, and other resources that a species or population required for survival. To understand the niche of the roseate spoonbill, Allen literally moved into it, outfitting a small, flat-bottomed boat as a floating laboratory and camping out for days on buggy, muddy islets in order to watch the birds at close range. He knew the elusive whooping crane would be an even tougher subject, and, as he remembered years later, "I wondered idly what poor, unsuspecting soul would someday be assigned the rugged task of making a full-scale study of them. I hadn't the slightest notion that it would be me!"

When the Audubon Society asked Allen to lead its whooping crane work at Aransas, he set about observing the rarest bird on the continent. During his first months at the refuge, he built a blind out of wire and canvas in the approximate shape of a bull, which worked admirably until a real bull approached the blind while Allen was crouched inside, ready to pick a fight with the intruder. (The live bull eventually wandered off, and a relieved Allen retreated to camp. He switched to better-defended blinds, but later allowed a *Life* magazine photographer to take his chances inside the faux bull.)

Allen also started spending summers in far northern Canada, surveying remote wetlands from small planes in hopes of finding the whoopers' nesting grounds. In 1955, after a nest had been spotted from the air, Allen and several colleagues attempted to reach the site on foot. On July 3, when they emerged from six weeks of searching, they were filthy, tired, and grinning. Deep in the nearly impassable backcountry of Wood Buffalo National Park—established three decades earlier to protect the last remaining bison herds in northern Canada—they had finally gotten a close look at a whooping crane nest, the first documented anywhere since the 1920s. "We had gazed at its features, and felt of it with our bare hands," Allen wrote in his memoir. "It is real and it is reasonably understandable. But most important of all, it is no longer unknown."

The team's success was celebrated throughout the continent, and a radio drama even imagined the whooping cranes' reaction to the discovery of their nesting grounds. George Archibald, the co-founder of the International Crane Foundation, remembers listening to the broadcast as a third-grader in rural Nova Scotia. "Now we will be shot, stuffed, and placed in a museum!" fretted a female crane. "No, dear, this is a safe place," a male crane assured her. "The wood bison survived here, and so will we."

Allen's triumph did not guarantee the survival of the species, however. Despite a determined public education campaign by the Audubon Society, hunters were still shooting whooping cranes—deliberately and by accident—as the birds traveled between Texas and Canada. Less than two weeks after Allen returned from the Canadian bogs, John Lynch, a biologist with the U.S. Fish and Wildlife Service in Louisiana, warned that the current flock of twenty-one adult cranes was not large enough to guarantee a future for the species.

Lynch had taught emergency survival to airmen while serving in the U.S. Navy during the Second World War, and he was trained to take decisive action. In a memo to his superiors, he argued that all of the surviving birds should be captured and bred. "And we ain't talking about semi-wild enclosures. In fact, we are not even talking about 'Wildlife Management,'" he wrote. "What is needed right now is an emergency, high-intensity period of 'Poultry Husbandry.'"

Allen knew the whooping crane was not out of danger, but he was not interested in poultry husbandry. Both sandhill and whooping cranes reproduce agonizingly slowly: they mate for life, and over a typical twenty-year reproductive span only the most fortunate pairs raise a dozen chicks to adulthood. The Audubon Society's attempts to breed whooping cranes at Aransas had been heartbreaking failures, resulting first in a clutch of infertile eggs and then in a single chick, Rusty, who lived for only three days. Even if captive whoopers could be encouraged to reproduce successfully, they would not be able to teach their chicks how to migrate—and in Allen's view, whooping cranes that could not migrate were not cranes in the fullest sense. He had spent years observing birds in their native surroundings, and the use of captive breeding to save a once free-flying species was, for him, a contradiction in terms.

The argument came to a head in late 1956, during a seven-hour meeting of Whooping Crane Project participants and supporters in Washington, DC. When Lynch claimed that the group already had a "backlog of information" on raising whooping cranes in captivity, Allen sharply disagreed, reminding his colleagues that the only whooping crane chick known to have been born in captivity had promptly died.

Both sides wanted to save the birds, and both feared that their opponents were about to accomplish the opposite. "Everybody was at each other's throats," Fish and Wildlife Service biologist Ray Erickson later recalled. Allen was outnumbered, and even his allies at Audubon would eventually agree to back captive breeding efforts. He remained unpersuaded, however, once writing to a colleague that if he had to choose between extinction and indefinite captivity for the whoopers, he would choose extinction.

On June 28, 1963, while at home in Florida, Allen had a heart attack and died on the way to the hospital. He was fifty-eight. The following year, his widow, Evelyn, received a visit from Stewart Udall, who presented her with a proclamation from President Johnson that designated three Florida islands as the Bob Allen Keys.

Udall shared Allen's distrust of heroic measures. "Is scarcity to be our only standard of value for wildlife?" he asked in a message to the attendees of the North American Wildlife and Natural Resources Conference in the spring of 1964. "There is something pathetic about a nearly-extinct creature. We can only partially applaud ourselves for halting it on the brink of extinction." Nonetheless, Udall concurred with Allen's colleagues that halting a species on the brink was better than watching it tip into oblivion.

• • • •

When Rachel Carson published *Silent Spring* in the fall of 1962, some species were already protected by national laws and policies, in the United States and elsewhere. The U.S. Migratory Bird Treaty Act, passed after Audubon activists' long campaign, had made it illegal to "pursue, hunt, take, capture, kill, attempt to take, capture or kill, possess, [or] offer for sale" any member of hundreds of bird species, including cranes. In the 1930s, following the suggestions of Leopold and others, President Franklin Roosevelt had not only established a raft of new wildlife refuges but signed the Pittman–Robertson Act, which extended a 10 percent tax on sporting arms and ammunition and distributed the revenue to states for wildlife conservation efforts.

Despite growing interest in conservation in North America and Europe, no country had yet passed a comprehensive law designed to prevent the extinction of species, and Udall returned from his trip to East Africa determined to do so. In January 1964, a little more than two months after the assassination of President Kennedy, Udall—who would serve as Interior Secretary until the end of the Johnson administration—announced the creation of the Committee on Rare and Endangered Wildlife Species, charged with advising the Department of the Interior on the "official designation of rare and endangered species and biotic communities," captive breeding, and "other" issues related to species conservation. Naturally, CREWS included the government scientists who had the most experience with endangered species: the whooping crane biologists who supported captive breeding.

The committee's recommendations to the Department of the Interior could have emphasized the importance of habitat, or of preventing species endangerment in the first place. The whooping crane biologists, however, had spent years facing down extinction; they were concerned with emergency cases, and convinced of the need for poultry husbandry. The CREWS recommendations, unsurprisingly, implied that captive breeding was as important to saving species as wildlife refuges and hunting regulations.

The committee also could have prioritized the "biotic communities" mentioned in its remit, but its members were not inclined to do so. While some prominent ecologists had begun to study entire ecosystems, the background and experience of the ecologists on the committee predisposed them to focus on single species. Ultimately, writes historian Johnny Winston, the committee members' narrow range of strategic and scientific perspectives "significantly constrained the shape of federal endangered-wildlife policy."

CREWS members were familiar with the IUCN Red Data Books—some had contributed to the series—and they readily accepted them as a model. In 1964, they assembled a preliminary version of what they called a "Red Book" for the United States, listing some sixty vertebrate species that they and their external scientific advisors agreed were at risk of extinction. The following year, when Udall sent draft endangered species legislation to Congress, it reflected CREWS members' concern with highly endangered species and their enthusiasm for captive breeding. The department's intention, Udall wrote, was to "conserve, protect, restore, and where necessary to establish wild populations, propagate selected species of native fish and wildlife . . . that are found to be threatened with extinction."

In 1966, Congress approved the Endangered Species Preservation Act, which required the Department of the Interior to continue keeping lists of endangered species and directed other government agencies to protect those species "insofar as is practicable." Those protections were expanded three years later, then significantly strengthened in 1973, when Congress passed the Endangered Species Act that remains in force today.

The law attracted wide support, and few detractors. The National Rifle Association testified on its behalf, and some of the most conservative members of the House and Senate backed it with little hesitation. Nobody gave much thought to its financial or political costs, because they expected both to be relatively small. Even the World Wildlife Fund estimated that most of the planet's seriously threatened species could be saved for one and a half million dollars per year.

Lawmakers and the press assumed that the Endangered Species Act was intended for large, well-known animals: the grizzly bear, the American alligator, the bald eagle, the whooping crane. Such species, after all, had so far defined the work of conservation. But the law was taxonomically generous. The U.S. "Red Book" had by then grown to a list of hundreds of species that included the Hawaiian hoary bat and the San Francisco garter snake, and a provision in the final legislation extended endangered species eligibility to plants. Ultimately, the only species deemed ineligible were insect pests and microorganisms. The law's definition of a species was later expanded to cover subspecies and, in the case of vertebrates, "distinct population segments"—groups of organisms identified as crucial to their species' persistence.

The Endangered Species Act also made some provisions for habitat. Late in the legislative process, the stated purposes of the act were expanded to include the protection of "ecosystems upon which endangered species and threatened species depend." The law permitted (and later required) the designation of "critical habitat," areas where federal agencies must consult with the Fish and Wildlife Service about their actions' effects on listed species.

And while the law's prohibition against the "taking" of endangered species was widely understood to apply to the outright killing of individual animals, not the destruction of habitat, the courts would decide otherwise. In 1978, after a University of Tennessee law professor and his students sued to stop the completion of a dam on the Little Tennessee River, arguing that the government-funded project would endanger a recently recognized species of perch called the snail darter, the U.S. Supreme Court ruled that yes, in fact, the law stipulated that federal projects must not destroy species or, by extension, their habitats.

Outraged that a three-inch-long fish might stop a multimillion-dollar dam, members of Congress responded to the ruling by creating the so-called "God Squad," a panel of experts empowered to create exceptions to the Endangered Species Act. When the God Squad, too, ruled against the dam, the Tennessee congressional delegation tucked an exemption into a spending bill. The dam was completed and the river valley flooded, but several transplanted populations of snail darters survived, and other populations were later discovered in nearby tributaries. The story of the snail darter has since become something of a fable in its own right, used—depending on who is telling it—to illustrate either the excesses or the successes of the law.

The Endangered Species Act had evolved significantly since its beginnings in CREWS, but the committee's emphasis on single species and emergency cases remained embedded in the law, and its influence is still felt. While only eleven of the domestic plant and animal species protected by the act have gone extinct, only forty-six have recovered to the point that they have been removed from the list. The rest—about seventeen hundred plant and animal species—remain on the list, and progress is slow. One analysis of a subset of species recovery plans devised under the law found that although the majority of species and populations had recovered on schedule or were on their way to doing so, their actual or projected time to full recovery averaged forty-one years.

The U.S. Endangered Species Act became a model for similar legislation around the world, and it remains one of the most far-reaching laws of its kind. Conservationists and environmentalists alike have spent decades defending it from political attacks, and rightly so. But while it is an indispensable bulwark against extinction, it often takes effect too late for species to make a speedy recovery—or any recovery at all. The most powerful species-protection law in the world is not powerful enough to fully protect other species from ourselves.

One of the strengths of the Endangered Species Act—and it does have many—is its ability to provide a conservation argument for an environmental strategy. The predicament of one well-known, well-loved species can illustrate the need to protect air, water, or land for many other forms of life.

When the bald eagle population in the contiguous United States was designated as endangered in 1967, the species was already protected from hunting throughout the country, thanks to the lobbying efforts spurred by Rosalie Edge almost forty years earlier. Even so, bald eagle numbers outside Alaska had continued to drop, and in 1963, the population in the lower forty-eight states hit a low of 417 nesting pairs. Though the endangered listing drew public attention to the bald eagle's troubles, the rescue of the species was already underway, and the effort had begun with another lawsuit over DDT spraying on Long Island.

On a spring afternoon in 1965, Carol Yannacone drove past Upper Yaphank Lake, a small millpond on the eastern end of the island. For her, it was a cherished spot; she had grown up swimming in the shallow water and exploring the nearby woods, and she liked to take her two young children there to play. When she passed the lake that afternoon, she was horrified to see that its surface was completely covered with dead fish. "When you know all the people around, you know the water, you know the fish, and then you see that—I still splutter, thinking about it," she told me.

Months later, Yannacone heard from a chemist at the local water authority that the fish kill had been caused by the driver of a county DDT spray truck, who had flushed the truck's contents into the pond. Though Yannacone hadn't read *Silent Spring*, she had studied biology in college and worked as a radioisotope technician at nearby Brookhaven National Laboratory. "I had heard some things about DDT, and none of them were flattering," she recalled.

When the county announced a new round of spraying, Yannacone persuaded her husband, Victor Yannacone Jr., a hard-driving lawyer with some experience in pollution litigation, to take up the issue. After filing a lawsuit on his wife's behalf against the Suffolk County Mosquito Control Commission, he called Charles Wurster, a biochemist at the nearby State University of New York at Stony Brook. He asked if Wurster knew anything about DDT, if he would support the lawsuit, and if any of his colleagues would lend their names to the effort. Wurster, as he remembered it, said yes, yes, and yes, and with the help of his evidence the Yannacones won a temporary injunction against the upcoming spraying.

   Like Robert Cushman Murphy and the other Long Island residents who had sued over local authorities' use of DDT a decade earlier, the Yannacones eventually lost in the courtroom but won beyond it. The mosquito commission stopped using DDT, and the case attracted so much public support that the Yannacones decided to expand their campaign. Borrowing a tactic from Rosalie Edge, the Yannacones and Wurster ambushed the board of the National Audubon Society by proposing, from the floor of the national convention, that the organization establish a legal fund and use it to pursue a ban on DDT. The membership supported the idea, but the startled board demurred, reluctant to commit to an extended legal fight. Wurster and the Yannacones, realizing they were on their own, recruited several sympathizers and incorporated themselves as the Environmental Defense Fund.

   Operating on a shoestring budget, EDF sued over DDT use in Michigan and Wisconsin, and in 1969 it challenged the continued use of DDT by the federal government. Two years later, an appeals court ordered the brand-new Environmental Protection Agency to essentially prohibit the use of DDT in the United States. The creation of the EPA had been partly inspired by *Silent Spring,* and President Richard Nixon had directed the agency to treat the environment as a "single, interrelated system." Yet after an eight-month agency hearing involving more than 125 witnesses—including scientists who testified to the effects of DDT on eagles, peregrine falcons, brown pelicans, and other birds—the presiding judge recommended that the agency allow DDT use to continue. EPA administrator William Ruckelshaus had the final say, however, and he issued his decision on June 14, 1972. While existing data on the human health effects of DDT were troubling but inconclusive, Ruckelshaus wrote, the evidence of the damage done to wildlife was convincing. "I am persuaded," he wrote, "that the long-range risk of continued use of DDT . . . is unacceptable and outweighs any benefits."

   The ban on DDT went into effect on December 31, 1972. Within a decade, nearly eighteen hundred pairs of bald eagles were nesting in the contiguous United States. In 2007, when there were ten thousand pairs, the population was removed from the endangered list, and its numbers have since continued to grow. The Environmental Defense Fund prospered as well, and today it employs hundreds of lawyers, economists, and other experts who help shape environmental policies worldwide.

Every time the Endangered Species Act celebrates a milestone birthday, the Fish and Wildlife Service touts the law as the savior of the national bird. While the listing of the bald eagle surely increased public—and official—support for a ban on DDT, the law's role in saving the species was secondary. The bald eagle's primary rescuers were those who cared about not only eagles but many other species—and who widened the concept of conservation to include us, too.

The whooping crane, like the bald eagle, was part of the "class of 1967"—the first group of birds, mammals, fish, reptiles, and amphibians protected under the Endangered Species Act. Unlike the bald eagle, the whooping crane remains in critical condition.

North America is now home to 667 free-flying whooping cranes— many times more than the all-time low of fifteen migratory birds recorded in the early 1940s, but not enough to consider the species recovered, or even safe from imminent extinction. The flock that migrates between Texas and Canada has grown to 504 birds, but it is still threatened by poachers, and its winter habitat is growing saltier due to the rise in sea levels.

Robert Allen feared that the captive breeding of whooping cranes would destroy the species, either by killing its few remaining representatives or erasing its collective ability to migrate. Neither fear has been realized, but even modest success has required ingenuity and stubbornness. In the mid-1970s, the International Crane Foundation in Wisconsin adopted Tex, a captive female whooper whose ancestry predated the species' midcentury crash. Foundation co-founder George Archibald, knowing that Tex carried a rare genetic legacy, was determined that she reproduce, but she was so used to human companionship that she refused to pair with another crane. Undeterred, Archibald offered himself as a substitute mate: he set up a desk outside Tex's enclosure so that he could be near her all day, and even learned to imitate the dramatic flapping and deep knee bends of the whooping crane mating dance. Archibald's four years of immersion in cranehood made it possible for Tex to be artificially inseminated, and eventually she produced a fertile egg and a weak but otherwise healthy chick, Gee Whiz.

By the 1980s, whooping cranes were regularly pairing and reproducing in captivity, but captive-raised whoopers were unprepared to travel the continent as their ancestors had. So when crane conservationists heard that a Canadian artist and inventor named Bill Lishman had taught a flock of captive-born geese to fly alongside his ultralight aircraft, they wondered if Lishman could also teach whoopers to migrate. Over the next decade, experiments by Lishman and collaborators with his organization, Operation Migration, found that flocks of

George Archibald and his "mate," Tex, on the grounds of the International Crane Foundation, 1982.

captive-raised geese, trumpeter swans, and sandhill cranes could all be persuaded to follow ultralights—and, after accompanying an aircraft south in the fall, would return north on their own in the spring, navigating independently for hundreds of miles.

Early on a frosty morning in October 2001, three bright yellow ultralights lifted off from Necedah National Wildlife Refuge in Wisconsin, eight young whooping cranes trailing behind them. The cranes had been raised by white-robed caretakers to prevent their imprinting on humans, and they had successfully completed several shorter practice flights. But no one knew whether they could manage the journey south, much less an unchaperoned return to Wisconsin.

During the forty-eight-day, twelve-hundred-mile trip to the Chassahowitzka National Wildlife Refuge on Florida's Gulf Coast, one bird died in a windstorm. Over the winter, bobcats killed two more. But in the spring, when five healthy, captive-raised whooping cranes returned to Wisconsin, it seemed to many that Operation Migration had performed a miracle.

The cranes, their costumed caretakers, and their airborne human flight instructors became famous worldwide. Each year, students of all ages followed along from their classrooms as a new batch of young birds embarked on its first flight south. By 2010, about one hundred whooping cranes were migrating between Florida and Wisconsin, and several crane pairs had produced chicks. Finally, it seemed, humans had figured out how to raise whooping cranes that knew how to be whooping cranes—and how to defy extinction.

Over the years, though, it became clear that during their captive upbringing, the Operation Migration cranes had missed some important lessons. In Florida, captive-raised adult cranes wandered away from nests, leaving their chicks exposed to predators, and showed only fitful interest in teaching their offspring to survive. During a decade of ultralight releases, only ten chicks born to the new migratory population lived long enough to fledge.

In early 2016 the Fish and Wildlife Service, which had worked with both the crane foundation and Operation Migration and holds legal authority over all of the captive-raised whooping cranes, ended the ultralight training flights. The white-robed caretakers were replaced with techniques designed to further minimize human contact with captive-born birds, such as the direct release of parent-reared chicks into free-roaming flocks. By the end of 2019, forty-two parent-reared chicks had been released into flocks. Twenty-three had died, with most killed by predators or in collisions with vehicles or power lines; five had reached breeding age; and two had produced chicks. None of those chicks had survived to fledge.

Why does one species recover while another struggles? The answers are rarely simple. The central threat to the bald eagle was removed before the species became as desperately endangered as the whooping crane; bald eagles are also more adaptable than whoopers, less choosy about their habitat and quicker to reproduce. The bald eagle's recovery in the lower forty-eight states did benefit from some captive breeding efforts, but the population was mostly able to rebound on its own. The bald eagle's rescue, in other words, was accomplished through an environmental strategy. The whooping crane's had to begin with poultry husbandry.

"This is a bird that cannot compromise or adjust its way of life to ours," Allen wrote in a 1952 monograph on the whooping crane. "Could not by its very nature; could not even if we had allowed it the opportunity, which we did not." While it is impossible to entirely untangle the effects of what Leopold called biotic violence from the generally more gradual selective pressures that have created and extinguished species for eons, Allen suspected that the whooping crane was on the wrong end of both. "We have singled out the Whooping Crane for survival for reasons that are peculiarly our own," he cautioned, "in the face of the possibility that Nature has already greased the skids that would lead to its ultimate destruction."

George Archibald is more optimistic than Allen about the whooper's persistence—and about humans' ability to ensure it. When Archibald and his graduate school classmate Ron Sauey founded the International Crane Foundation, many of the world's crane species were in such straits that captive breeding was the foundation's primary strategy; a Russian collaborator once sent Sauey home from Moscow with five Siberian crane eggs nestled in a suitcase packed with hot-water bottles. (One chick hatched en route, and was named Aeroflot.) Now, with some species out of immediate danger of extinction, Archibald and his foundation colleagues can spend more of their time on what Udall called the total environment—on the habitat that cranes and other species need to flourish.

While the foundation's whooping crane program remains focused on captive breeding, it is also protecting and restoring whooper habitat in Texas and elsewhere—and continuing the hunter education efforts that the National Audubon Society began in the 1940s. Archibald has witnessed enough crane comebacks to believe that whoopers, too, will one day cross the threshold to relative safety, and he wants to be ready. "When we get it right," he told me with a beatific smile, "we're going to have a lot of them back."

Today, captive breeding is probably the best-known way to save a species, thanks in part to its unforgettable milestones: a newly released California condor, soaring free over the Vermilion Cliffs of northern Arizona; a baby panda rolling around with her nursery mates; an intrepid flock of whoopers sailing behind an ultralight. But the thrill and drama of these captive breeding efforts, as Robert Allen well knew, owes a great deal to their high stakes and real risks. Even dedicated practitioners of captive breeding acknowledge that it is a last resort—the intensive care of conservation.

And like intensive care, captive breeding is unrelentingly expensive. Organizations such as the International Crane Foundation are supported in large part by private donations; public money for chancy, decades-long conservation efforts is finite, to say the least. The whooping crane breeding program at the Patuxent Wildlife Research Center, established in the mid-1960s with eggs collected from nests in Wood Buffalo National Park, moved most of its cranes to private facilities after it lost its government funding in late 2017. The North American Model of Wildlife Conservation, established before the IUCN turned the conservation movement's attention toward highly endangered species, was conceived by Leopold and others as a way to protect game and other species still numerous enough to survive outside captivity—not to support decades of captive breeding.

When Stewart Udall left the Department of the Interior, he knew that threats to species—visible and invisible, direct and indirect—were accelerating, and that the endangered species law he had helped develop was only one step toward protecting the "total American environment." Further progress required the principles of ecology to be more consistently applied to policy—and that, he believed, required ecologists to apply themselves to politics.

In a 1971 essay titled "The Ivory Tower Is Under Siege," Udall claimed that he would rather see scientists "err on the side of activism and occasional 'hyperbole'" than stifle their responses to environmental crises. "We need their indignation at the fouling of man's habitat and the misuse of his resources as much as we need their ideas for better solutions," he wrote. "We need scientists who dare to stretch their minds and relate their expertise to the whole human enterprise."

Indignant scientists were already descending from the ivory tower. "Qualified biologists are beginning to forge a discipline in that turbulent and vital area where biology meets the social sciences and humanities," biologist David Ehrenfeld wrote in 1970. The practitioners of the new discipline would call themselves conservation biologists, and they would find the territory vital, but turbulent indeed.

Chapter Seven

THE SCIENTISTS WHO ESCAPED THE TOWER

n 1978, on a September evening in San Diego, California, an up-and-coming biology professor named Michael Soulé addressed an audience at the San Diego Zoo's open-air Safari Park. As the assembled scientists and conservation professionals dined within earshot of the zoo's resident pride of lions, Soulé, then in his early forties, warned that the world was on the verge of the largest species extinction since the disappearance of the dinosaurs. He proposed the founding of a field that would apply lessons from wildlife biology, ecology, evolutionary biology, and other specialties to pressing conservation problems. Only by overcoming the widening gap between theory and practice, he said, could science provide the tools that humans needed to do right by other species.

Soulé's listeners, who were attending the optimistically named First International Conference on Conservation Biology, were well aware of the growing threat of human-caused extinctions, and most welcomed a new means of addressing it. The prospective new discipline did attract skepticism: Some wildlife biologists dismissed it as a trendy version of their own field—only with more pretensions and fewer data. Some ecologists doubted that evolutionary theory had much to contribute to conservation. Some researchers thought that a "crisis discipline," as

Soulé would come to call conservation biology, sounded too political, or too much like an excuse for shoddy science.

These detractors were relatively few, though. Many researchers were confident that a crisis discipline could produce rigorous science, and welcomed the chance to collaborate on questions relevant to conservation. Students who were studying ecology or evolutionary biology because they cared about the fate of other species saw conservation biology as a chance to make their science matter. Private and public funders were drawn to the prospect of science-based conservation solutions.

In the years after the publication of *Silent Spring*, ecological principles such as the food web became so well known that the environmental movement was sometimes called the "ecology movement," yet few professional ecologists had tried to influence environmental or conservation politics. Conservation biology promised to change that.

"Disciplines are not logical constructs," Soulé wrote. "They are social crystallizations which occur when a group of people agree that association and discourse serve their interests." Conservation biology, in other words, appeared when enough people decided to call themselves conservation biologists. What they didn't know was that their new discipline would reveal the limits of biology.

"And think, also, what it would mean if even one-half the men and women who earn their daily bread in the field of zoology and nature-study should elect to make this cause their own!" stormed William Hornaday during a lecture at the Yale School of Forestry in 1914. Despite his repeated appeals for help, he lamented, "fully ninety per cent of the zoologists of America stick closely to their desk-work, soaring after the infinite and diving after the unfathomable, but never spending a dollar or lifting an active finger on the firing-line in defense of wild life." As allies in the cause of conservation, he concluded, zoologists were "hopelessly sodden and apathetic."

Hornaday was, as always, prone to overstatement, but it was true that many zoologists and naturalists had long kept their distance from conservation causes. Like other scientists, they were generally inclined to avoid "meddling with Divinity, Metaphysics, Moralls, Politics, Grammar, Rhetorick, or Logick," as the British polymath Robert Hooke had put it in 1663 during the formation of the Royal Society. Public expressions of political opinions, scientists believed, would cloud their view of reality—threatening the quality of their science and, perhaps just as importantly, their reputations as dispassionate scientists.

In 1863, Alfred Russel Wallace, whose evolutionary insights startled Charles Darwin into publishing his long-gestating theory, warned his colleagues of "the extinction of the numerous forms of life which the progress of cultivation invariably entails." Instead of encouraging naturalists to speak up for these species, however, Wallace advised them to make "perfect collections" of animal and plant specimens for posterity. Twenty years later, a similar inclination would spur Hornaday to hunt down the surviving plains bison.

Toward the end of the nineteenth century, some professional ornithologists on both sides of the Atlantic joined the campaign against the plume trade, striving to preserve bird species not only in "perfect collections" but in life. Others, however, abstained from "meddling with Politics"—partly out of fear that conservation laws would limit their own ability to kill and preserve birds for research.

One of the first ecologists to lift a finger on the conservation firing line was Victor Shelford, a professor at the University of Illinois who studied animal and plant communities. Shelford helped found the Ecological Society of America in 1915, and soon afterward dedicated a committee within the new society to the preservation of "natural conditions" in parks and public forests. Introspective and intensely private, Shelford nevertheless pursued his mission with zeal, enlisting the support of Gifford Pinchot and pressuring both the National Park Service and the U.S. Forest Service to take science more seriously. In a 1921 article in the journal *Science*, zoologist Francis Sumner, the committee's co-chair, put out a call for reinforcements. "It is my hope," he wrote, "that more of our leaders in science will be aroused to the necessity of becoming also leaders in the conservation movement." The ecological society's executive committee, however, feared that Shelford's frequent buttonholing of legislators and bureaucrats reflected badly on the organization, and they limited and eventually revoked the powers of the preservation committee.

Shelford was not deterred. "Human society, which supports research, will hold scientific men and the societies which they constitute responsible for failure to urge the application of their knowledge," he wrote in *Science* in 1944. "No scientific society devoted to research should fail in fulfilling this obligation."

In 1946, Shelford and his supporters formed an independent organization called the Ecologists Union, which they dedicated to the protection of "biotic communities." Shelford did not live to see the crystallization of conservation biology, but he did have the satisfaction of witnessing the rapid early growth of the Ecologists' Union. (Aldo Leopold mailed in his own membership dues at the end of 1947, four months before his death and shortly before beginning an abbreviated term as president of the Ecological Society of America.) Today, the organization employs more than four hundred scientists, works in seventy-nine countries, and has so far protected more than 125 million acres of land worldwide. Since 1951, it has been known as the Nature Conservancy.

In the 1990s, Soulé retired from his university career in California and moved to rural western Colorado, where he resided until his death in June 2020. He and his wife, June, lived in a compact, one-story house with a wide-angle view of the Rocky Mountain foothills, and for several years during the final decade of his life, I was their neighbor, our houses separated by an irrigated pasture and a short stretch of unpaved road. Soulé was rail-thin, with a monklike aspect: bald head, angular features, and penetrating gaze. Though he could be quick to anger, he had a disarming capacity for delight, flashing a small, mischievous smile at a joke or a bird at the window.

He cried easily, a tendency he found embarrassing but occasionally

useful, and he choked up over tragedies as small as the death of a single turtle and as large as mass extinction. What really bothered him, he
often said, was not the prospect of death but that of the end of birth—
the end of evolution, the end of possibility.

On a shelf in his home office was a simple brass microscope, a long-
ago gift from his mother. In the 1940s and 1950s, when Soulé was
growing up on the Pacific coast near San Diego, California, he used the
microscope to inspect leaves and drops of water—minutiae gathered
during his adventures in the thick stands of sagebrush and scrub oak
near his house. "I would just lie down and look up through the sumac
and feel very much at home," he told me.

Though there were no other naturalists in his immediate family,
his mother and stepfather—Soulé's father died when he was two—
encouraged his interests, not only giving him a microscope but allowing him to keep lizards and even, at one point, a badger in his room.
(The badger proved to be a rambunctious roommate and was returned
to the desert.) From an early age, he was driven by both scientific curiosity and a deep affection for other forms of life. "When I'm in a place
with many creatures, I feel good, and when there are no creatures
around, I feel bad," he once reflected. "It's not rational."

When Soulé was a teenager, a high school teacher advised him to
join the junior naturalist program at a local museum. There, Soulé
and his friends studied birds and lizards the way other kids studied
baseball statistics. They went to the coast to investigate tidepools and
collect abalone. When they were old enough to drive, they crossed the
border into Mexico to look for unfamiliar species of reptiles (during
one return trip, he and his companions successfully hid an illicit bottle
of rum under a bag of rattlesnakes). From the car, they competed to
be the first to recognize a species of plant or animal and shout out its
Linnaean name.

Despite these joyful expeditions, Soulé came to see himself as
something of an outsider. Part of the reason, he speculated, may have
been that he was a member of one of the only Jewish families in

an overwhelmingly Gentile neighborhood, during an era when real estate agents in at least one nearby community refused to sell homes to Jews. Whatever its sources, the feeling lasted, and became something of a professional advantage; while all scientists are trained to maintain a critical distance from their work, Soulé was also able to cast an analytical eye on the scientific establishment. Like Leopold, he would join many respected institutions—and work to subvert them from the inside.

During the early years of his career, Soulé was preoccupied not with saving the world but with learning how it worked. He was especially fascinated by anatomy—by how organisms were put together, and by the physical differences among individuals and species. For his doctoral dissertation at Stanford University, he studied populations of side-blotched lizards on islands in the Gulf of California, finding that the lizards on larger islands closer to the mainland were more variable in appearance than lizards on smaller, more isolated islands. Using an early method of genetic analysis, Soulé showed that the physical variation within populations of lizards—in characteristics such as the number of scales around their right eyes and on one of their hind toes—corresponded with their level of genetic variation.

Soulé was exploring the theory of island biogeography, which had been proposed a few years earlier by the mathematically inclined ecologist Robert MacArthur and his colleague Edward Wilson, later known to the public as E. O. Wilson. Building on Wallace's long-ago observations on the islands of Southeast Asia, MacArthur and Wilson proposed that the number of species on an island was determined by the island's size and its rate of immigration. Because larger islands closer to the mainland would experience more immigration, more speciation, and less extinction, they should support a greater number of species overall. (Wilson and his graduate student Daniel Simberloff tested this theory by poisoning all of the resident spiders and insects on seven islets in the Florida Keys, then watching as the islets were repopulated.)

Soulé's work suggested that this theory of species diversity also

applied to genetic diversity. Both kinds of diversity would prove important to the protection of species, on islands and off.

While Soulé was counting lizard scales in the Gulf of California in
the early 1960s, the public's sense of planetary crisis was escalating.
Global vulnerabilities to famine and disease, exposed by the Second
World War and vividly enumerated by Fairfield Osborn and William
Vogt in their twin 1948 jeremiads, seemed to many in North America
and Europe to be greater than ever.

Soulé's sense of crisis was shared by biologist Paul Ehrlich, his graduate adviser at Stanford. Ehrlich had been fascinated by butterflies
since his childhood in suburban New Jersey; in high school, he worked
for the curator of the butterfly collection at the American Museum of
Natural History in New York in exchange for discarded specimens.
From his own research, Ehrlich knew that butterfly populations tended
to grow until, hit by food or space shortages or an outbreak of disease,
they crashed.

Ehrlich had read both Osborn and Vogt as a college student, and the
looming resource shortages they described convinced him—and many
others—that humanity was headed for a butterfly-like crash. In 1968,
at the invitation of David Brower, then the director of the Sierra Club,
Ehrlich and his wife, Anne, co-wrote a book on the subject. It became
not only a bestseller but a lasting metaphor: *The Population Bomb*.

Ehrlich thought the book's title was alarmist—he initially wanted to
call it *Population, Resources, and Environment*—but he was, in Stewart
Udall's words, willing to "err on the side of . . . occasional 'hyperbole.'"
The opening scene of the book, in which the Ehrlichs recount a taxi
ride through the crowded streets of Delhi, is pure Vogt: "The streets
seemed alive with people. People eating, people washing, people sleeping. People visiting, arguing, and screaming . . . People defecating and
urinating. People clinging to buses. People herding animals. People,
people, people, people."

The Population Bomb drew little attention at first, but Ehrlich was

Paul Ehrlich and Johnny Carson on the set of *The Tonight Show*, 1979.

unusual among scientists for his comfort in the spotlight, and his frank, highly quotable warnings soon earned the book a gigantic audience. "It's really very simple, Johnny," Ehrlich said to Johnny Carson during one of his many appearances on *The Tonight Show*. More people meant less food per person, and more starvation. It was, Ehrlich said, "already too late to avoid famines that will kill millions."

Ehrlich's booming voice amplified the long-running debate about the significance of human numbers to the fate of life on earth. Like the Huxley brothers, Ehrlich generally accepted the predictions of eighteenth-century cleric Thomas Robert Malthus: unless otherwise controlled, human societies would grow exponentially until checked by famine. Critics of Malthus, meanwhile, maintained that modern humans could escape catastrophe—and reduce their impact on other species—through advances in technology.

In 1981, the economist Julian Simon, frustrated by Ehrlich's omnipresence in the media, publicly challenged the biologist and his allies to a bet on the future costs of raw materials. If prices fell or remained the same, Simon would win on behalf of prosperity. If they rose, Ehrlich's side would win on behalf of scarcity. "How about it, doomsayers and catastrophists? First come, first served," Simon wrote in *Social Science Quarterly*. Ehrlich, along with his close friends and collaborators John Holdren and John Harte, eagerly accepted, betting that the prices of chromium, copper, nickel, tin, and tungsten would rise by the end of the decade. Simon won, earning $576.07—a sum based on the combined drop in the metals' prices.

If the men had agreed on different commodities, or a different time period, the wager could have easily tipped toward Ehrlich, Holdren, and Harte. But Simon was right to question Malthus. When the Ehrlichs published *The Population Bomb*, new high-yielding crop varieties were already being planted worldwide. Between 1960 and 2000, the "Green Revolution" in agriculture tripled wheat harvests and more than doubled rice and corn harvests in poorer countries. While these monocrops and the synthetic fertilizers used to grow them created new environmental problems—and concentrated political and economic control of food supplies—their bounty temporarily suspended the Malthusian cycle.

Which is not to say that our species, or any other, is immune to limits—only that those limits are not as fixed as we sometimes assume. Carrying capacity, the ecological concept that preoccupied Vogt and Leopold, is rarely quantifiable or stable, especially when it comes to humans. The capacity of any given environment to feed its human residents changes over time, and can be radically affected by circumstances. Even though the Green Revolution significantly increased the global food supply, for instance, devastating regional food shortages are still caused by war, corruption, climate extremes, and inequalities of wealth and power. Likewise, the impact of any given human on an environment varies with affluence, access to technology, and individual choices.

In 2009, earth systems scientist Johan Rockström and his colleagues at the Stockholm Resilience Centre proposed the concept of "planetary boundaries," an alternative to the notion of global carrying capacity. Instead of using food supplies to estimate the earth's human capacity, Rockström and his colleagues studied the fundamental processes and resources that make the planet habitable by humans—the water cycle, the nitrogen cycle, the carbon cycle, and so on—and estimated their tolerances for disruption. The researchers then used those tolerances to define what they called a "safe operating space for humanity." In an article for the journal *Nature*, they warned that human use of fossil fuels had already disrupted the global carbon cycle, destabilizing the climate; that the synthetic fertilizers popularized by the Green Revolution had overloaded the nitrogen cycle, polluting waterways and killing marine life; and that human-caused species extinctions were testing the resilience of ecosystems.

Ehrlich and his fellow "doomsayers and catastrophists," as Simon called them, had a prescient respect for planetary boundaries, and they recognized that humanity's approach to those boundaries is hastened by our increasing numbers. But they underestimated the mitigating potential of technology—and of individual human choices, especially concerning reproduction. "There is no reason to expect that the millions of decisions about family size made by couples in their own interest will automatically control population for the benefit of society," sociologist Kingsley Davis wrote in *Science* in 1967. "On the contrary, there are good reasons to think they will not do so."

At the time, the Ehrlichs heartily agreed. "The story in the UDCs [underdeveloped countries, then a common term] is depressingly the same everywhere—people *want* large families," they wrote in the first edition of *The Population Bomb*. "They *want* families of a size that will keep the population growing." (Malthus, for his part, concluded that "preventive checks," including the "moral restraint" required to postpone sex, were more prevalent in Europe than "in the more uncivilized parts of the world.")

Over the past half century, these claims have been consistently contradicted by the results of voluntary family planning programs, which have reduced birth rates and improved the overall health of women and children worldwide—especially when combined with greater access to education for girls. Countries including Bangladesh, Kenya, Indonesia, and Iran have reduced average family sizes and population growth rates by expanding—and officially condoning—reproductive health services. "When wisdom dictates that you do not need more children," the Iranian ayatollah Ali Khamenei declared in 1996, "a vasectomy is permissible."

Meanwhile, coercive population control campaigns like that conducted in India in the mid-1970s—during which more than eight million women and men underwent sterilization—and the one-child policy in China violated basic human freedoms, with ghastly unintended consequences, and were less effective over time than voluntary measures. In 1994, the United Nations convened representatives of 179 countries in Cairo for the International Conference on Population and Development. By acclamation, the representatives affirmed that "women's ability to control their own fertility" was essential to the success of any population program.

The United Nations now projects that global population growth will essentially level off by the year 2100. The surest way to accelerate this trend, research suggests, would be to invest in family planning programs and girls' education in Pakistan, Nigeria, the Democratic Republic of the Congo, Tanzania, Ethiopia, and Angola—the six countries expected to account for more than half the world's population growth through the end of the century.

Looking back, Ehrlich says that *The Population Bomb* should have stressed the importance of women's rights, and explicitly rejected racism. "The thing we *know* works best is improving gender equity and racial equity," he told me. "If you want to do something about population, give full rights and opportunities to women, including access to abortion." He also wishes that he and Anne had emphasized the need to reduce not only the overall number of humans but the rate of resource consumption by the rich.

Ehrlich, however, stands by their warning: the planet has a finite capacity to produce food and support people, and we ignore its limits at our peril. In 2015, an update to the planetary boundaries framework confirmed that humans are still marching toward, and past, those limits.

A few months after the publication of *The Population Bomb*, Ehrlich's neo-Malthusian arguments were seconded by biologist Garrett Hardin in his *Science* essay "The Tragedy of the Commons." Hardin contended that humans, when left to their own devices, inevitably exploit all available resources, through both breakneck breeding and unrestrained competition with one another. While Ehrlich flirted with coercive measures, warning that they might be necessary if voluntary means failed, Hardin embraced them. "The freedom to breed is intolerable," he wrote, and the only solution was "mutual coercion, mutually agreed upon." Like Vogt—and, to a lesser extent, Ehrlich—before him, Hardin publicly discouraged the provision of food aid to some poorer countries: "The less provident and less able will multiply at the expense of the abler and more provident, bringing eventual ruin upon all who share in the commons." He compared wealthy nations to lifeboats that could not accept more passengers without sinking.

The tragedy of the commons, however, is not inevitable. Hardin's critics, notably the Nobel Prize–winning political scientist Elinor Ostrom, would establish that at many scales and in many situations, people are capable of cooperating for mutual benefit. (Ostrom and her colleagues argued that what Hardin called a tragedy is really a drama, in that it is a story whose ending can go either way.) But the tragedy of the commons, like the population bomb, is an uncannily enduring metaphor, and both continue to exert an outsized influence not only on conservation biology but on the conservation movement as a whole.

To his graduate students at Stanford, Ehrlich was a hands-off mentor— "He never knew exactly what I did my dissertation on," Soulé told me— but he made sure that his students had the practical support and the intellectual skills they needed to pursue their research. "Paul taught me how to think like a scientist," Soulé said. "He wouldn't let anyone get away with sloppy thinking." Ehrlich challenged his students to venture outside their disciplines, and imbued them with his sense of urgency. Unchecked human population growth, he predicted, would sooner or later bring about not only widespread famine but the mass extinction of other species.

When Soulé finished his graduate work, he landed a professorship at the University of California, San Diego, and returned to his hometown. The city was growing, and many of the canyons where he had chased lizards as a boy were filling up with roads and houses. To him, the changes symbolized those underway throughout the world, and strengthened his conviction that Ehrlich was right about extinction.

In the mid-1970s, Soulé was contacted by Otto Frankel, an Austrian wheat geneticist who spent much of his career arguing for the importance of preserving genetic diversity in crop plants, both in seed banks and in active cultivation. Frankel was concerned—rightly, as it turned out—that the genetically uniform Green Revolution crops would prove vulnerable to disease and pest outbreaks and less adaptable to changing conditions, and would imperil what he called the "treasuries of variation" found in more localized crop varieties. Frankel thought that genetic diversity in nondomesticated species might be similarly important to their survival, and he invited Soulé to collaborate.

There was, and is, a cultural divide between theoretically inclined scientists like Soulé and "applied" scientists who, like Frankel, seek solutions to real-world problems in agriculture, wildlife management, and other fields. Some applied scientists dismiss theorists as impractical snobs; some theorists dismiss applied scientists as unimaginative plodders. Many of the most innovative scientists, however, have bridged the gap. Leopold, trained in applied science, didn't hesitate to incorporate ecological theory into his work, and Soulé likewise welcomed the chance to collaborate with Frankel. "I wanted to be famous, of course, like every scientist does," Soulé told me, chuckling. "But

I didn't care whether it was applied or theoretical—I never had that prejudice."

When Soulé was a teenager, learning the names of as many plants and animals as he could find, he got some advice from an older naturalist: "When in doubt, count." By the time Soulé began working with Frankel, he had both an intellectual and gut-level understanding of the importance of counting to species survival—whether the counted were people, or islands, or the number of scales on lizard toes. Species survive when enough individuals have enough resources—food, water, space—to maintain the variation needed to adapt and persist. But how much was enough? That question, Soulé knew, could occupy an entire scientific field.

Say there were such a field, he thought. What would it be called? Population biology, Ehrlich's specialty, asked similar questions but to different ends. One day, while Soulé was in the shower, it came to him. He would call it conservation biology.

While Soulé's speech at the San Diego Zoo Safari Park announced the beginning of a new discipline, it was also, for him, a farewell; the following year, he resigned from his tenured professorship, gave away his house, and moved with his wife and their three young children to the Zen Center of Los Angeles.

Soulé had always had what he simply called a "spiritual side," a feeling of oneness with the rest of life, and he was powerfully drawn to Zen Buddhism. Once at the center, though, he had difficulty adjusting to the smog and noise of Los Angeles, and after several months found that he was restless. "I realized that I wasn't meant to become a full-time religious person, that I was meant to be a full-time scientist," he told me. His wife, meanwhile, wanted to commit even more deeply to Zen practice. Eventually, Soulé and his wife separated, dividing their parenting responsibilities, and he left the center and returned to San Diego. (Soulé remained a serious student of Zen Buddhism for the rest of his life, and his former wife, Jan Chozen Bays, is now the co-abbot of a Buddhist monastery in Oregon.)

Soulé had deliberately closed the door on his academic career, and he had to pry it back open. For about a year, he worked as an unpaid assistant in the University of California, San Diego, laboratory of Michael Gilpin, a biologist he had hired a decade earlier. In 1984, Soulé accepted a half-time position at the University of Michigan, and the next year he convened the Second International Conference on Conservation Biology—almost seven years after the first. On the final day of the conference, the attendees voted to form a society that would support their nascent field, and Soulé was charged with bringing it to life.

After that, the discipline crystallized rapidly. By the time the Society for Conservation Biology was six years old, it had five thousand member scientists; the Ecological Society of America, founded seven decades earlier, had only six thousand five hundred. Universities had launched at least sixteen academic programs in conservation biology.

The growth of conservation biology was aided by the almost simultaneous—and even more auspicious—debut of another conservation concept. In 1985, Harvard entomologist E. O. Wilson, the co-author of the theory of island biogeography, published a journal article called "The Biological Diversity Crisis." Wilson, by that time, was not only a widely recognized scientist but an accomplished author and communicator. Like Ehrlich, he was willing to spend time in the spotlight; like Soulé, he was willing to publicly profess his love for other species. As an entomologist, he was also exquisitely aware that the nearly two million species described by science were only a small sample of the variety of life on earth.

In his article, Wilson argued that a complete catalog of species was urgently needed, not only because of its value to science but because its source material was under threat. "Under the best of conditions," he wrote, echoing Soulé's rallying cry in San Diego, "the reduction of diversity seems destined to approach that of the great natural catastrophes at the end of the Paleozoic and Mesozoic Eras, in other words, the most extreme for sixty-five million years."

While the public, and even many scientists, gravitated toward large, furry vertebrates, Wilson was a student of ant communities, and he knew that some of the most wondrous forms of life were also the tiniest. Even seemingly unremarkable species were important because of their particular place in the grand web of life—their contribution to the overall diversity of species. "It might still be argued that to know one kind of beetle is to know them all, or at least enough to get by," Wilson wrote. "But a species is not like a molecule in a cloud of molecules. It is a unique population of organisms, the terminus of a lineage that split off thousands or even millions of years ago."

In September 1986, the Smithsonian Institution and the National Academy of Sciences convened the National Forum on BioDiversity in Washington, DC. (Biologist Walter Rosen suggested the contraction, which Wilson at first rejected as "too glitzy" but quickly came to recognize as powerfully succinct.) Dozens of prominent biologists, economists, philosophers, and professional conservationists spoke on topics ranging from rainforest ecology to the artificial insemination of endangered animals. Press coverage of the proceedings was ample, and on the conference's final evening, a discussion among Wilson, Ehrlich, Joan Martin-Brown of the United Nations Environment Programme, and other well-known scientists was broadcast to thousands of students at universities throughout North America.

"Biodiversity is humanity's most valuable and also its least appreciated resource up till now," Wilson told the virtual audience. "The biodiversity crisis—by which we mean the accelerating extinction of species and genetic strains around the world—is a real crisis, and of all the losses we may suffer in the years immediately ahead, remember: this is the only one that's irreversible." (When Ehrlich held forth on the need to limit human population growth, Wilson lent his teasing support: "Paul is known as a bit of a partisan on this issue, and I think another voice should be added on the matter. I will do so by saying that I think he is one hundred percent correct.")

The importance of biodiversity—it soon lost its capital D—was quickly accepted by the public, and in 1992 the signing of the Convention on Biological Diversity at the Rio Earth Summit enshrined the concept in international governance. Even the inertia-prone scientific establishment embraced the term. A search of one comprehensive index of scientific publications turned up zero references to "biological diversity" or "biodiversity" in 1981, and only seven references in 1982. In 2005, articles in the same index referred to biodiversity nearly four thousand times.

The concept of biodiversity, as historian Timothy Farnham writes, was a powerful "mediator, promoter, and catalyst." Biodiversity encompassed the proliferating concerns of the conservation and environmental movements; hunters, wilderness activists, animal welfare campaigners, and advocates for clean air and water could all find something to like about biodiversity. Ecologists, meanwhile, had set aside their longstanding attachment to the "balance of nature," having recognized that systems did not tend toward stability but were instead shaped by disturbance. Biodiversity, which referred to genetic and ecosystem diversity as well as species diversity, called attention to the dynamism of ecological systems and provided ecologists of all kinds with a new paradigm, turning onetime students of balance into students of diversity.

In a 1985 paper called "What Is Conservation Biology?" Soulé enumerated what he called the "normative postulates," or common beliefs, of conservation biologists. Diversity of organisms is good (with the corollary that the untimely extinction of populations and species is bad). Ecological complexity is good. Evolution is good. Biotic diversity has intrinsic value. While Soulé didn't mention Leopold, the postulates were his land ethic for conservation biology: A thing is right when it tends to preserve biological diversity, ecological complexity, and the evolutionary process. It is wrong when it leads to untimely extinctions.

While Soulé proposed that these postulates guide the work of conservation biology and serve to measure its success, he acknowledged that the goodness of diversity "cannot be tested or proven." Since Darwin, scientists had recognized that individual variety within species and populations allowed those groups to adapt and evolve, and work by Frankel, Soulé, and others was showing that genetic variety mattered, too. Wilson, in his paper, had argued that the global diversity of species was also important. While some species might provide the raw material for valuable crops or pharmaceuticals, he wrote, even those with no practical value to humans were unique repositories of evolutionary history. But how important was species diversity to the survival of particular forests, prairies, and other living systems? Neither Wilson nor anyone else could say for sure.

Scientists had long suspected that the variety of species within an ecosystem contributed more than simply the sum of its parts. Darwin himself had reported that a field planted with distantly related grasses was more productive than one planted with a single species. Charles Elton, the ecologist who proposed the concept of the food chain, hazarded in 1958 that relatively simple communities of plants and animals were more vulnerable to what his friend Leopold had called "biotic violence." Decades of research, however, had failed to pin down a connection between species diversity and what researchers began to call "ecosystem function."

The sudden popularity of the biodiversity concept drew a new generation of researchers to the lingering questions beneath it. Ecologists have since found that, in many cases, species play complementary roles within ecosystems, leading to not only greater overall productivity but greater resistance to disease and other disturbances. Yet there are many kinds of biological diversity, and their influences on living systems are varied, difficult to measure, and only partly understood. "More biodiversity" is not a universal prescription for conservation—and some conservationists argue that the concept of biodiversity is too broad and vague to be much of a prescription at all. Biodiversity matters—but to some extent, its goodness is still an article of faith.

The first generation of conservation biologists spent most of its time trying to pinpoint how much was enough: how many individuals did a population need in order to survive, and how large did a reserve need to be in order to protect a species? These questions were related to the theory of island biogeography, which suggested that diversity within terrestrial reserves, like that of islands, would be determined by the size and isolation of the reserve. But they were prompted by real-world conservation problems.

In the United States, the environmental laws passed in the 1970s— the National Environmental Policy Act, the Clean Air Act, the Clean Water Act, the Endangered Species Act, and others—led to a slew of new rules for government agencies, and many of the rules called for scientific guidance. In 1979, when the National Forest Management Act led to a regulation requiring "viable populations" of vertebrate species to be maintained in the national forests, U.S. Forest Service ecologist Hal Salwasser knew that the size of those populations would not be easy to define. In the early 1980s, he asked Soulé for help.

A few years earlier, Soulé and geneticist Ian Franklin had used studies of fruit fly populations and livestock breeding to come up with a rule of thumb for population viability: fifty individuals were required for short-term survival, and five hundred to maintain the diversity required for long-term survival. Soulé knew this rule was far less precise than it sounded—Gilpin remembered Soulé's frustration when a group of Australian scientists called to ask if there was any hope for a population of forty-eight parrots—and Salwasser's request prompted him to come up with a more nuanced approach.

Over the next year, Soulé and Gilpin developed what they called population viability analysis, or PVA, which uses a particular population's size, habitat characteristics, and genetic variability to predict its extinction risk over time. After Salwasser saw Soulé and Gilpin present their analysis, he exclaimed that they finally "had it"—namely, a way to estimate the amount of old-growth forest required by the northern spotted owl, a species of increasing concern to Forest Service biologists in the northwestern United States. (While population viability analysis would not head off an epic political battle over the owl, it would replace researchers' back-of-the envelope acreage estimates with more credible numbers.)

The more Soulé studied the requirements for species survival, however, the more complexities he discovered. In the 1980s, he and several students undertook a study of the quail, roadrunners, wrens, and other bird species that depend on the sage- and oak-filled canyons surrounding San Diego. Each of the thirty-seven canyons they studied was, in effect, an isolated island in a sea of development, and, as predicted by the theory of island biogeography, they found that smaller, longer-isolated canyons had fewer species of resident birds. These effects were magnified by what Soulé and his students called "mesopredator release": as coyotes, which only occasionally prey on birds, abandoned the smaller, more isolated canyons, they were replaced by a flood of raccoons, foxes, and domestic cats—all of which made quick work of the birds. Cats, their stamina buoyed by a reliable supply of store-bought food, appeared to be the most lethal hunters of all.

The safe operating space for a species, it turned out, was defined not only by the size and isolation of reserves but by the behavior and appetites of the surrounding residents. Ecologists found that the effectiveness of reserves was also influenced by what they called "edge effects"—the heightened vulnerability of reserve edges to disturbances of all kinds, from storms to domesticated cat incursions. By 2008, these variables had multiplied to the point that, as tropical ecologist William Laurance observed, Wilson and MacArthur's "elegant and important" theory of island biogeography appeared "simplistic to the point of being cartoonish."

While the science of reserve design grew more sophisticated, the politics of parks and protected areas were becoming both more enlightened and more complicated. In the years after Julian Huxley campaigned to establish the IUCN, conservationists had expanded their ambitions, working to protect more species in more places, and parks were their tool of choice. Between 1969 and 1980, the total area of what the United Nations classified as "national parks and equivalent reserves" more than doubled, from about 420 million acres to nearly a billion acres, about half the size of Brazil. By 1980, more than 10 percent of the land area of several African countries—including Botswana, Tanzania, Zimbabwe, and Rwanda—lay within protected areas. Though the ocean had received relatively little attention during the early decades of the conservation movement, eloquent advocates such as William Beebe, Rachel Carson, and scuba pioneer Jacques Cousteau had inspired public interest in marine conservation, and by the mid-1970s more than one hundred marine protected areas had been formally established worldwide—including the iconic Great Barrier Reef Marine Park in Australia, created in 1975.

Many of these parks were sources of national pride, and some were economic boons for their human neighbors. Others were reminiscent of the colonial parks of the late 1800s and early 1900s: established by national governments in cooperation with international conservation groups, often with little regard for the people living in and around them. Project Tiger, an Indian government initiative launched in 1973 and supported by the IUCN and the World Wildlife Fund, created a series of reserves that staved off the imminent extinction of the Bengal tiger—but also displaced thousands of people. As the IUCN and WWF learned when their representatives tried to keep Maasai herders out of parks and reserves in East Africa, the disruption of local livelihoods could leave parks surrounded by hostile humans.

In the late 1960s and early 1970s, warnings such as those expressed in *The Population Bomb* and a widely read 1972 report called *The Limits to Growth*—which predicted that global per-capita food supplies would peak around the year 2020—convinced many politicians and policymakers that opportunities for economic development were finite. Poorer nations, seeking a more equitable piece of the circumscribed pie, demanded greater control over their natural resources. At the United Nations Conference on the Human Environment in Stockholm in 1972, Julius Nyerere, the Tanzanian leader who had resisted Julian Huxley's "Declaration of a State of Emergency," led a concerted call for declarations against colonialism and apartheid. Parks, already suspect as a colonial hangover, began to look like an obstacle to national growth.

Within the international conservation movement, tensions rose. Some argued that conservation and economic development were not necessarily at odds, and that conservation principles could and should be used to promote sound development. Others insisted that the conservation movement should stay focused on creating parks and saving species, leaving human concerns to others. IUCN chief ecologist Raymond Dasmann, who had studied deer population dynamics under Leopold's son Starker at the University of California, Berkeley, argued that the top-down approach of most international conservation groups ensured that neither development nor conservation efforts would benefit the people who needed them most.

These internal battles came to a head in 1975, at the IUCN General Assembly in Kinshasa, Zaire. The meeting, which was hosted by Zaire's dictatorial president Mombutu Sese Seko and held in a heavily armed compound, was fraught from the start, and so fractious that participants later referred to it as the "Night of the Long Knives." The IUCN leadership was ousted, beginning a series of changes that further committed the organization to the "conservation for development" approach. Dasmann, for his part, successfully championed a resolution affirming the importance of protecting indigenous livelihoods in the management of protected areas.

Within five years, what came to be called "sustainable development"—development that "meets the needs of the present without compromising the ability of future generations to meet their own needs," according to one prominent definition—became the central goal of the international conservation establishment. The IUCN World Conservation Strategy, published in 1980, described conservationists as having "allowed themselves to be seen as resisting all development" and called for "global strategies both for development and for conservation of nature."

In theory, sustainable development was practiced by rich and poor nations, minimizing environmental damage by all. Former Norwegian prime minister Gro Harlem Brundtland, chair of the United Nations World Commission on Environment and Development, emphasized in 1987 that "the 'environment' is where we all live; and 'development' is what we all do in attempting to improve our lot within that abode." In reality, the responsibility for developing sustainably fell to poorer nations and those who sought to assist them. The definition of development, as Brundtland wrote, was too often reduced to "what poor nations should do to become richer."

In pursuit of sustainable development, conservation groups such as the Environmental Defense Fund persuaded the World Bank and other international lending agencies to require environmental reviews of dams, highways, and other loan-funded projects. In the 1990s, after the international Earth Summit in Rio, lenders stepped up their funding of "integrated conservation and development projects," initiatives intended to benefit both parks and people by reducing poverty in and around protected areas.

Despite some successes, progress was slow and failures were common. International conservation groups gained some influence over large development projects, but many projects were fundamentally irreconcilable with conservation and yielded few benefits for any species, including the majority of humans. (Many were also vehemently opposed by grassroots conservation groups, indigenous people's groups, or both.) Conservationists who had believed in the promise of sustainable development began to doubt that conservation and development could be compatible, much less mutually beneficial.

Soulé, in his 1995 essay "The Social Siege of Nature," warned his colleagues away from what he described as "the profane grail of Sustainable Development—the odd delusion of having your cake and eating it too." By the late 1990s, some prominent conservationists were calling for a renewed emphasis on securing the borders of parks and other protected areas, an approach which by then was often disparaged as "fortress conservation." Tropical ecologist John Terborgh, who had spent much of his career in the Peruvian rainforest, suggested that an internationally financed corps of armed rangers be employed to patrol park boundaries. "Parks cannot be maintained without order and discipline," he wrote, "but order and discipline are antithetical to many of the corrupt, freewheeling societies of the developing world."

In 2003, Wildlife Conservation Society scientist Kent Redford, a veteran of many IUCN committees and assemblies, published a poignant editorial in the journal *Conservation Biology*. Redford and his co-author, M. Sanjayan, a Nature Conservancy scientist who had studied under Soulé, called for a more honest accounting of the costs of development—and conservation. "To change the fate of the world, conservation biology must provide scenarios balancing human well-being and a world rich in nature, as well as the scientific basis for evaluating the trade-offs inherent in these scenarios," they wrote. "But this is not enough; we must also convince society to choose options that increase the conservation of nature in all its splendor."

Win-win solutions had proved elusive. To protect biodiversity—to provide other species with the resources they needed to adapt, survive, and thrive—conservationists, including conservation biologists, had to persuade some of their fellow humans to make some sacrifices, at least in the short term. But how?

While the international conservation movement chased the dream of sustainable development, Soulé had been pursuing a vision of his own. His continuing research, and his work with the Society for Conservation Biology, had repeatedly reminded him of the importance of connectivity—among patches of habitat, academic disciplines, and researchers and activists.

Soulé and his colleagues were also aware of the ecological importance of what they called "apex consumers." In grasslands, oceans, rainforests, and other ecosystems, researchers had found that species near the top of the biotic pyramid—bison, wildebeest, sea stars, jaguars—exert grazing or predation pressure on the layers below them, maintaining diversity by heading off population explosions of any single species. Much like the mountain lions on the Kaibab Plateau, the predators in these systems appear to prevent Malthusian catastrophe.

One of North America's top predators, the gray wolf, had been nearly eliminated from the contiguous United States by hunting and by the government trapping and poisoning campaigns opposed by Joseph Grinnell and Rosalie Edge. The gray wolf gained endangered status along with the whooping crane and the bald eagle as part of the "class of 1967," and in the late 1970s, wolves from Canada began to wander into northwestern Montana. As their numbers grew, Soulé and others started to envision a network of habitat that would allow wolves and other predators to travel along the Rocky Mountains from Canada to Mexico, providing the space their populations needed to survive as well as the opportunity to reclaim their role at the top of the biotic pyramid. The notion of connecting "cores" of habitat with "corridors" navigable by mobile species was not new, but no one had seriously suggested doing so on a continental scale.

Soulé, who by that time was a professor in the environmental studies department at the University of California, Santa Cruz, convened a group of biologists and activists that included Douglas Tompkins, cofounder of The North Face, an outdoor clothing company, and Dave Foreman, co-founder of Earth First!, a radical environmental group. In 1991, they started the organization now known as the Wildlands Network, and began to map existing reserves and potential habitat linkages throughout the United States. For the conservation biologists involved, it was a giant step further into politics. While they had advised policymakers on conservation questions, they didn't typically propose policy themselves, and this policy proposal was eye-poppingly ambitious.

"You hold in your hands, I sincerely believe, one of the most important documents in conservation history," Foreman wrote in the introduction to the group's manifesto, published in the magazine *Wild Earth* in 1992. "What we seek is a path that leads to beauty, abundance, wholeness, and wildness. We look for the big outside instead of empire, we seek wolf tracks instead of gold, we crave life rather than death."

While the proposal promised that jobs "will be created, not lost," and that land "will be given freely, not taken," it was a broadside on behalf of biodiversity, long on ecological and evolutionary principles and short on economics and social concerns. "To stem the disappearance of wildlife and wilderness we must allow the recovery of whole ecosystems and landscapes in every region of North America," Soulé and his co-authors wrote. "Allowing these systems to recover requires a long-term master plan."

The politics surrounding protected areas and species protection were by that time almost as bitter in the United States as they had become in parts of Africa, Asia, and South America. For more than a century, conservationists in North America and Europe had used their considerable powers of persuasion to pass laws restricting human behavior. They had often faced determined opposition, but they had succeeded in shielding bison from market hunters, egrets from plume traders, songbirds from overenthusiastic ornithologists and natural-history collectors, and eagles from DDT—with few lasting costs, and obvious benefits for human health and recreation.

Wildlife professionals had also convinced recreational hunters to accept limits and fees in exchange for continued access to game—the arrangement established by the North American Model of Wildlife Conservation. Many of the professionals who upheld the model, however, were hunters themselves, and few followed Leopold's instruction to "recognize conservation as one integral whole, of which game restoration is only a part." Rather than persuading the non-hunting public to support the "jointly formulated and jointly financed program for protection of all wildlife" that Leopold envisioned, they relied on hunters to pay for conservation, with the result that most species protection efforts by state wildlife agencies were (and still are) financed by hunting license fees and taxes on hunting equipment.

When the Endangered Species Act became law in 1973, it abruptly changed the terms of conservation in the United States—and, as other countries passed similar laws, around the world. While conservationists had convinced North Americans and Europeans that the survival of bison and magnificent birds was well worth the sacrifice of luxuries such as buffalo rugs and feathered hats, the persistence of less charismatic species elicited less sympathy. In the 1980s, when it became clear that some endangered species protections—like those proposed for the northern spotted owl—carried costs that would not be covered by hunting license fees, the backlash against conservation was lacerating.

Promises of long-term benefits for the economy, and the planet, did little to placate timber workers who feared their jobs would be affected by spotted owl protections, or ranchers whose grazing privileges on public land were limited by desert tortoise protections. Resentment of the federal government, endemic in the rural United States even before Theodore Roosevelt created the Forest Service, was quickly exploited by the corporations and politicians whose short-term interests were also threatened by conservation.

Soulé was well aware of these tensions, and his contribution to the Wildlands Network proposal advised supporters to practice a "politics of patience." But for those who believed that conservation had already gone too far, the proposal itself was incendiary. Opponents drew up a fake map that depicted the reserve network in blood red, covering the nation like a rash, and copies appeared in hardware stores and gas stations throughout the rural United States—sometimes accompanied by dark warnings of a global conservation conspiracy, one supposedly created by Julian Huxley and carried on by the IUCN.

The Wildlands Network proposal also fired the imaginations of conservationists, many of whom were weary of documenting declines and longed for a positive vision—even one that might take centuries to enact. Paul Ehrlich and E. O. Wilson endorsed the proposal, as did other prominent conservationists and conservation biologists, and conservation-minded private foundations supported its dissemination. (Some conservation biologists were more critical, citing uncertainties about whether highly mobile species would use habitat corridors.)

In the mid-1990s, Soulé left California and moved to rural Colorado, where he devoted himself to promoting the "cores, corridors, and carnivores" vision of the Wildlands Network. In 2008, he heard that a wolf, or wolves, might have taken up residence at a large guest ranch not far from where he lived. Wolves had been driven out of Colorado seventy years earlier, and while individual wolves had wandered into the state since the species had been reintroduced in Yellowstone National Park in 1995, there were no known resident wolves in Colorado. But Cristina Eisenberg, a biologist hired to survey the guest ranch, had found droppings and tracks whose characteristics strongly reminded her of those left by the wolves she studied in Montana. Once, while driving across the ranch, she had seen a dark, wolf-like shape flash across a field. Someone else had reported hearing a distant howl.

None of the evidence was anywhere close to conclusive, and Eisenberg and her field assistants even half-jokingly avoided the word "wolf," calling the unidentified animal or animals only "the visitors from the north." But it was enough to convince the ranch owner, Paul Vahldiek Jr., to send the droppings to a laboratory in California for genetic analysis.

For Soulé, even the distant possibility that a wolf had come back to stay in the southern Rocky Mountains was enough to draw him to the ranch, and on repeat visits he struck up a conversation with the owner about making room on the property for wolves.

Vahldiek, a wealthy businessman and hunting enthusiast from Texas, initially scoffed at Soulé's ideas, but after attending a Wildlands Network meeting and seeing his three-hundred-square-mile property on a map of potential habitat linkages, he became cautiously enthusiastic about the possibilities, both for his business and the landscape. Maybe a resident pack of wolves would attract visitors; maybe it would reduce the numbers of elk, and by doing so revive the aspen trees that the elk herds nibbled down each year. Vahldiek, who controlled his own ecosystem from the top down, became an unlikely but valuable ally.

During the months that Vahldiek was awaiting the results of the genetic analysis, I joined Soulé, John Terborgh, and a group of other conservation biologists for a walk along the shale and sandstone cliffs above the ranch. The winter day was cold and clear, and the scientists were relaxed, talking quietly. They knew that the closest living wolf was most likely hundreds of miles away. Still, there was a trace of tension in the air. The return of an apex species is a grand experiment in ecosystem reconstruction, and the scientists—most of whom had spent decades witnessing ecosystem demolition—couldn't help but hope that it was already underway.

Soulé, his head exposed to the elements, had wrapped his bony frame in a heavy camouflage coat. (He had recently taken up hunting, both for sport and—because his recreation seemed to require a cerebral element—as an investigation of his personal ethics.) He stepped off the road to inspect a set of tracks. Shifting his scrutiny from the snow to the horizon and back, he allowed himself to imagine that a wolf was just out of sight. "It feels wonderful," he said with a grin. "I'm not frightened at all."

The "visitors from the north" on Vahldiek's ranch, as it turned out, were probably not visitors, and almost certainly not wolves. The DNA results suggested they were coyotes. But wolves continued to show up in Colorado, and in early 2020, after a flurry of reported wolf sightings, state wildlife officials confirmed that a pack of at least six wolves was living north of Vahldiek's ranch. Wolves from the northern Rocky Mountains have also ventured into the Pacific Northwest, founding more than four dozen resident packs in Washington and Oregon. In Europe, where humans are more and more concentrated in cities, wolves are striking

Michael Soulé looking for wolf tracks at the High Lonesome Ranch, 2010.

out from remnant populations in southern Europe and finding their way back to the northern European countryside. Like bison, returning wolves are inevitably met with both delight and consternation, polarized reactions that fuel polarized politics.

The Wildlands Network, its rhetoric now tempered, continues to advocate for habitat corridors between North American reserves. Other conservationists and conservation groups have floated even more ambitious plans. E. O. Wilson, in his 2016 book *Half-Earth,* argued that "only by committing half of the planet's surface to nature can we hope to save the immensity of life-forms that compose it." By setting aside the "largest reserves possible for nature," Wilson wrote, and stipulating that they are "allowed to exist unharmed," humanity could halt the decline of more than 80 percent of the species on earth. Without such an effort, Wilson predicted, humans will lose much of the living planet that sustains them.

Like the scientists of the Wildlands Network, Wilson is highly sensitive to the complexity of ecological relationships, and he devoted much of *Half-Earth* to descriptions of it; each species, he wrote, "lives intimately with other species—prey, predator, inside and outside symbionts, engineers of soil and vegetation." He spent far less time examining the human complexities of his proposal, instead predicting—with the optimism of Julian Simon—that the combined effects of the free market and high technology will shrink our collective footprint and make room for massive reserves.

He may be right. But to realize anything resembling his grand vision—to "convince society to choose options that increase the conservation of nature in all its splendor," as Redford and Sanjayan wrote—conservationists need to pay a lot more attention to human complexity.

Conservationists and environmentalists have acquired a reputation, in some circles, for being anti-human—for putting the interests of other species ahead of their own. Critics of conservation biology have periodically suggested that the field polish its image, and advance its cause, by reorienting itself toward human needs—prioritizing places, and species, that are most beneficial to humans. Because conservation needs human supporters to succeed, the argument goes, the work of conservation biologists should support the needs of humans.

Soulé and many other conservation biologists have reacted to these critiques with alarm and anger. If conservation biologists don't speak for biodiversity, they say, who will? "People aren't endangered," Soulé told me. "Biodiversity is."

On one level, the criticism of the field is undeserved. An analysis of all of the research published in major ecology and conservation biology journals between 2000 and 2014—a total of more than thirty-two thousand papers—found that many of its proposals for protecting biodiversity already paid close attention to human needs. But the critiques attract attention in part because they are on to something. Conservation biology does have a blind spot, and it's the same one that has afflicted the conservation movement since the plight of the plains bison shocked Hornaday into action. The problem isn't inattention to human needs, but inattention to human complexity.

Conservation biologists, given their training, tend to assume that humans are much like deer, or butterflies, or any other species. This is true up to a point, as Darwin memorably informed us, and both the conservation and environmental movements have usefully recalled humans to their dependence on clean air and fertile soil. But while we may not be any *better* than, say, jellyfish—and there are plenty of situations in which jellyfish are more capable than humans—we are *different* from jellyfish, and from every other species on earth. We are distinguished by our wildly complex societies, technologies, and means of expression, and by our multilayered individual variety. Another distinction, suggests philosopher Christine Korsgaard, is that we are aware of ourselves as members of a species. While many animals recognize others of their own kind, and are aware of their membership in pairs, packs, or flocks, humans regularly refer to all of humanity as "we." And because we place the stories we tell about ourselves within the stories we tell about our species, we are able to imagine humanity, in all its variety, taking something like collective action—including action on behalf of other species.

The assumption that only particular kinds of humans are distinctive—that only a subset of the "we" is different from other animals—underlies some of the darkest chapters of the conservation movement. It persuaded Hornaday and his colleagues to display Ota Benga, a member of their own species, in a cage. It convinced many early-twentieth-century conservationists to embrace the pseudoscience of eugenics, and a few to support the "applied biology" of the Third Reich. It led William Vogt to argue that food aid to poorer countries would disrupt "natural" controls on human numbers; it led Earth First! co-founder Dave Foreman, in a notorious interview in the wake of the

Ethiopian famine of the mid-1980s, to say that "the worst thing we could do in Ethiopia is to give aid—the best thing would be to just let nature seek its own balance, to let the people there just starve." (Foreman belatedly apologized.) The same assumption is central to the incoherent manifestos of today's "ecofascists," who use it to justify violence in the name of conservation.

More often, conservationists take a simplistic view of humanity as a whole, which is just self-defeating. It leads them to assume that the population bomb and the tragedy of the commons are inevitable, long after the inevitability of both has been disproven; to decide that Leopold's vision of humans as plain citizens of the planet is impossible, and that the only way to save other species is to wall them off from ourselves; to imagine political change as either a spontaneous grassroots uprising or a global fiat. It leads them to approach the concepts of wilderness and biodiversity not as tools for protecting life's variety but as evocations of a lost frontier—subscribing to what historian William Cronon describes as "the illusion that we can escape the troubles of the world in which our past has ensnared us."

"When in doubt, count" is excellent advice, on the whole. But numbers are neither the beginning nor the end of the story, especially when it comes to humans. Our singular awareness of ourselves as members of a species, and our ability to sort other organisms into types, allows us to both perceive the diversity of life on earth and conceive of protecting it. As Soulé found when he studied lizard populations in the Gulf of California, however, variation *within* species is also important—and to ignore the social, political, and economic variation within the human species is to ignore the immense differences in culpability, and vulnerability, that exist among us today.

The field of conservation biology is now more than forty years old, and its normative postulates have held up pretty well. The mission of conservation biology, and conservation in general, can still be summarized as the protection of biological diversity, ecological complexity, and the evolutionary process—in short, the preservation of possibility.

Perhaps conservation biology does need one more normative postulate—a reminder that humans are capable of protecting the rest of life from ourselves. Like the "goodness" of biological diversity, the truth of this statement can't be tested or proven. In fact, we're all bombarded with evidence to the contrary—which is why conservation biologists, and the rest of us, could use an article of faith in humanity. To dismiss our complexity, our ability to be both constructive and destructive, is to give up on the whole project of species conservation and, indeed, on the human project itself. And the outcome of that needs neither testing nor proving.

Conservation biologists have become experts in documenting the threats to other species, and in estimating how much space, food, and other resources are needed to combat them. The rest of the work of conservation—designing nuanced policies and protections that can attract the popular support they need to achieve their purpose— requires expertise in human complexity. Whether conservation biology broadens into conservation science, or branches into fields like conservation sociology and conservation agriculture, its practice can't be left only to the biologists.

"Biodiversity conservation is a human endeavor," environmental policy scholar Michael Mascia and several colleagues observed in a 2003 *Conservation Biology* editorial, "initiated by humans, designed by humans, and intended to modify human behavior to achieve a socially desired objective." Soulé acknowledged as much in 1986, when he called on conservation biologists to "ask for the help of professionals in other disciplines," including anthropologists, sociologists, and community development workers. "Otherwise," he wrote, "we risk espousing solutions that are theoretically robust, but socially and politically naive."

Leopold, among others, recognized this decades ago. Shortly before his death, he proposed a faculty position in ecological economics at the University of Wisconsin, "based on the premise that industrialization is now bringing on a worldwide conflict between economics and conservation." His favored candidate was his friend William Vogt.

"One of the anomalies of modern ecology is that it is the creation of two groups, each of which seems barely aware of the existence of the other," Leopold reflected in the mid-1930s. Sociologists, economists, and historians studied the human community "almost as if it were a separate entity," and biologists studied plants and animals as if they could be isolated from politics and other human concerns. "The inevitable fusion of these two lines of thought," Leopold wrote, "will, perhaps, constitute the outstanding advance of the present century."

Well into the next century, conservation biology has not yet managed this advance, but its practitioners are diving deeper into what biologist David Ehrenfeld once described as "that turbulent and vital area where biology meets the social sciences and humanities." In the late 1990s, ecologists and economists began an effort to estimate the value of what they called "ecosystem services" and "natural capital," attempting to quantify the myriad benefits—from clean water to mental health—that ecosystems deliver to humans. In the early 2000s, the Millennium Ecosystem Assessment, an international collaboration among more than a thousand researchers, used ecosystem services as a measure of planetary health.

For Leopold, quantifying the public benefits of conservation was only a first step in a deeper rethinking of the relationship between economics and conservation. He would surely agree with modern critics of the ecosystem services concept who argue that any price put on the priceless inevitably falls short. He would readily admit that even the most thorough valuation of an ecosystem is unlikely to tempt those interested only in profits; as conservationists have learned from experience, conservation measures rarely recoup their costs in the short term. But he would also be likely to agree with Gretchen Daily, the director of the Center for Conservation Biology at Stanford University and a leading proponent of ecosystem valuation, who points out that even a rough approximation of conservation's value to humans is better than its current default value in major policy decisions—which is zero.

In 2019, after more than thirteen hundred conservation biologists, social scientists, government officials, and others gathered in Kuala Lumpur for the annual international conference of the Society for Conservation Biology, they acknowledged these and other developments in the field with a declaration that called, in part, for "appropriate social and economic incentives for biodiversity conservation." The society's small corps of social scientists is growing, and Mascia, who served as president of the society from 2017 to 2019, was the first social scientist to hold the post. He studies the human factors affecting the long-term management of parks and reserves worldwide—and he wrote his doctoral dissertation on the applications of Elinor Ostrom's work to marine reserves.

Though biologists routinely argue that biodiversity benefits everyone, social scientists know that the costs and benefits of conservation are unevenly distributed. In many cases—and all parts of the world—the poor carry the burdens of conservation, while the wealthy enjoy most of the ecosystem services. These inequalities animate much of the popular antipathy toward laws like the Endangered Species Act and proposals for new, larger parks and reserves. And perhaps nowhere are the gaps more striking, or more entwined with the history of conservation, than in southern Africa.

Chapter Eight

THE RHINO AND
THE COMMONS

E ven in the antipodal winter, northwestern Namibia is brutally hot
and short on shade, and the cluster of battered trailers and cabins
that marks the old diamond prospecting camp of Wêreldsend—a
variation on the Afrikaans for "World's End"—offers little protection
from the punishing elements. Or from anything else, for that matter.
In early April of 1982, when Garth Owen-Smith arrived in Wêreldsend
to establish a field base for the brand-new Namibia Wildlife Trust, he
was abruptly reminded that the weather is only one of the dangers in
the desert.

During his first night in camp, Owen-Smith was startled awake
by what sounded like a pig being slaughtered. When he cautiously
approached the commotion, he learned that a young male lion had
padded into camp, sunk his teeth into a sleeping bull terrier, and
disappeared into the shadows, the dog still squealing in his jaws.
Owen-Smith and a companion managed to shoot the lion—the terrier, miraculously, survived the ordeal—but when they drove around
Wêreldsend to check for more intruders, their headlights illuminated
another half dozen lions, all hiding in the grass less than one hundred yards from camp. Owen-Smith knew that while lions would attack
humans if provoked, they much preferred to avoid people; the animals

laying siege to Wêreldsend had been driven to extremes by two years of deep drought.

Over the following weeks and months, Owen-Smith learned that the region's people and wildlife were in terrible straits. Though the drought had begun to ease, it had killed tens of thousands of domestic cattle— an estimated 90 percent of the region's herds—depriving many of the local Himba and Herero people of not only milk and meat but their social standing and sense of worth. A number of men, humiliated by their losses, had died by suicide.

Desperate to support themselves and their families, many people were ignoring the strict hunting limits the government had imposed during the drought, killing antelope and other game "for the pot." Others—including some government officials—were poaching elephants and black rhinos and selling tusks and horns to commercial traders. Once-healthy populations of elephants and rhinos were devastated. In a report to the IUCN Species Survival Commission in 1980, South African conservationist Clive Walker had estimated that there were fewer than eighty elephants and fifteen rhinos left in the region. As Owen-Smith explored the desert's rust-red basalt ridges and dry riverbeds, he sometimes encountered more carcasses than living animals.

Owen-Smith hoped to persuade people to stop poaching, but he knew that many had no alternative. He started by visiting the settlements where people were doing relatively well, sustained by their surviving cattle and some remaining game. As he got to know them and their circumstances, he asked if they valued the local wildlife.

Most people said no. Lions and leopards were eating the last of their livestock and threatening people, too. Elephants were trampling their crops and raiding their gardens. Now, the government was punishing them for hunting a few antelope for food. They bore all of the costs of conservation, and realized none of the benefits. What value did wildlife hold for them, they asked, other than as a source of illicit protein and occasional payoffs from commercial smugglers?

As in many other places in Africa, wildlife—especially the large, sometimes troublesome species that conservation biologists call apex consumers—had in recent decades come to be seen in the region as government property, protected for wealthy foreigners. For some locals, already alienated from other species, desperation had led to exploitation, which led to further alienation. During his first months of work with the Namibia Wildlife Trust, Owen-Smith often despaired of reversing this downward spiral.

One day he found himself listening, with increasing irritation, to an elderly man complain about a group of elephant bulls. The bulls were always drinking from the reservoir near his house, the man said, and they kept stomping on his belongings and breaking branches off his fig trees. Owen-Smith, who had heard the man's litany before, said shortly that if the elephants were such a problem, the only solution was to shoot them. The man paused, absorbing what Owen-Smith had said. His response, when it came, was indignant: "No one should shoot one of *my* elephants!"

Maybe, Owen-Smith thought, there was hope after all.

When I visited the Kunene Region of northwestern Namibia in late 2019, another drought was underway, one that many predicted would soon outstrip the dry spell of the 1980s. As I traveled the rough back roads with Owen-Smith and, later, guide and translator Edison Kasupi, we heard that hungry predators were once again attacking livestock. We passed the bloated carcass of a mountain zebra, dead of thirst or hunger.

But the desert was also full of life. One morning, when Kasupi and I left camp early and drove up the dry bed of the Hoanib River, we found the trees draped in sea mist, blown in overnight through a notch in the coastal mountains. A giraffe stood almost motionless under the leafy canopy, neck and head forming a graceful vertical line as she stretched upward to reach her breakfast. As we rounded the next bend, an adult elephant emerged from the trees, ears flapping in the fog. Four more followed. When we got a clear view of the group, we saw that the adults

A group of female elephants and their young walking along a dry riverbed in the Kunene Region.

were accompanied by a three-month-old baby—who was grazing on the brush while sheltered under her mother's belly—and an even smaller baby, barely a month old, who was trying not to step on his own trunk as he staggered up a sandy bank.

We heard there were lions in the area, too, but we didn't see them. The only hints of their presence, as we continued upstream, were the two community game guards camped on opposite sides of the riverbank, posted there to quietly discourage the lions from moving closer to livestock and humans.

Much had changed for the better in Kunene since Owen-Smith's eventful night in Wêreldsend nearly forty years earlier, and while he was reflexively loath to claim credit, the changes had at least something to do with him. He first fell in love with the region in 1967, after dropping out of college in his native South Africa and taking a job at a copper and lead mine in Namibia—a former German colony that was then known as South West Africa and controlled by South Africa. Invited by an acquaintance to visit the northwestern corner of the territory, he took an impromptu vacation from the pit bottom and traveled toward the Atlantic coast.

For Owen-Smith, the open horizons were an exhilarating contrast to the dark, cockroach-infested tunnels of his day job. He was struck by the severe beauty of the landscape, and by the abundance of life that managed to survive in it. He was also fascinated by the local customs: the brightly colored, elaborately petticoated "big dresses," inspired by the gowns of turn-of-the-century German missionaries, worn by many Herero women; the sweet-smelling *otjize* paste, made from butterfat and ochre pigment, that the semi-nomadic Himba used to protect their skin and sculpt their hair.

When Owen-Smith returned to work at the mine, he lasted less than a month before he quit, loaded his few belongings into his pickup truck, and started looking for a way back to the northwestern desert. After an ill-fated road-biking adventure across the Kalahari and two months working at an oil refinery near Durban, he was offered a position as an agricultural supervisor in northwestern South West Africa with the South African agency then called the Department of Bantu Administration and Development. He accepted with enthusiasm.

Owen-Smith became one of the region's handful of white residents, and one of a very few who regularly traveled the back roads and isolated valleys of its communally held lands. His tall, gaunt figure was soon widely known, and his full beard and habit of climbing to the nearest high point earned him the Herero nickname *Ondwezu jongombo*, the goat ram. (He considered himself lucky, since one of his colleagues was known as *Ombindo korambi*, the thin pig.)

Though he kept strategically quiet about his opposition to his employer's apartheid policies, some minor acts of resistance put Owen-Smith at odds with his supervisors, and after a little more than two

years he was forced to move back to South Africa. When he returned in
1982 the region was still beautiful, but its circumstances had changed,
and all of its species were suffering—not only scarred by drought but
shadowed by the protracted and bloody civil war in nearby Angola,
and divided by an escalating power struggle between South Africa and
South West Africa.

Owen-Smith was, in some ways, an unlikely candidate for a career
in conservation. He had no formal training in science, and because he
far preferred the isolation of Wêreldsend to the large conferences fre-
quented by the conservation establishment, he had few connections in
the international organizations that dominated the conservation movement in Africa. But he was stubbornly committed to the desert, and
with the help of mentors had taught himself the skills he needed to
serve it, gaining fluency in subjects ranging from car repair to animal
tracking to ecological theory. He had spent part of his decade-plus away
from South West Africa helping to manage a cattle ranch in present-
day Zimbabwe, testing conservation-oriented grazing practices in the
midst of a multipronged guerrilla war against the country's white-
minority government.

He also understood, as Michael Mascia and other social scientists
would later point out in *Conservation Biology*, that conservation is a
human endeavor, designed by humans to achieve a shared goal. To that
end, Owen-Smith liked to say, his most useful tools were his ears.

In the mid-1970s, when the international conservation movement
began to prioritize sustainable development over the creation and
maintenance of parks and reserves, many established conservation
groups traded one top-down strategy for another. Instead of negotiating
with national governments over appropriate reserve boundaries, they
began negotiating with international lenders over the size and shape of
development projects.

But a new generation of conservationists was emerging, not only
in Europe and North America but in Africa, Asia, and Latin Amer-

ica. Many agreed with Raymond Dasmann, the IUCN ecologist who had spoken up for indigenous rights during the organization's chaotic meeting in Kinshasa in 1975: In order to protect other species, the conservation movement needed to work with the humans who lived alongside them—and often depended on them as a source of food.

Despite conservation's long association with recreational hunters, both conservationists and environmentalists had largely ignored subsistence hunters as a potential constituency, and some had openly condemned them. Early conservationists like William Hornaday and Madison Grant believed subsistence hunters were beneath them; early environmentalists like Rachel Carson thought hunting was cruel. But conservationists were recognizing that support from subsistence hunters and farmers was often critical to the success of parks and reserves. Without it, even the best-funded, most thoroughly patrolled parks were little more than lines on maps, vulnerable to incursions by human neighbors. And unless those neighbors had a reliable means of feeding their own families, through hunting or otherwise, they could not be expected to tolerate—much less protect—the large, potentially destructive and dangerous species whose populations required more space than any single reserve could provide.

In a bid to both support local livelihoods and reduce pressure on other species, some conservationists began collaborating with these communities to strengthen existing wildlife management practices. Over the next decade, they developed a set of strategies known as community-based natural resource management or community-based conservation, aimed at rebuilding the local authority at the foundation of the conservation pyramid.

The earliest community-based conservation initiatives were relatively simple cost-sharing arrangements between park administrators and nearby communities. In 1977, for instance, Maasai leaders in southern Kenya—where some Maasai had been so outraged by restrictions on hunting and cattle grazing that they killed rhinos in protest—reached a short-lived but influential accord with Amboseli National Park in which they were compensated for the cost of accommodating park wildlife on traditional pastureland.

Central to such initiatives was the concept of "wildlife utilization" or "sustainable utilization"—the practice of "harvesting" financial or caloric benefits from other species while managing their populations for long-term survival. Julian Huxley, among others, had supported utilization as a key conservation strategy in postcolonial Africa. Within the international conservation movement, however, utilization had suffered from not only the historical snobbery toward subsistence hunting but the widespread belief that, given a chance, local people would inevitably enact what Garrett Hardin, in 1968, had called the tragedy of the commons.

When a resource is shared, Hardin argued, it is only rational for each individual to "utilize" that resource as much as possible—with predictable results. "Ruin is the destination toward which all men rush, each pursuing his own best interest in a society that believes in the freedom of the commons," he wrote. "Freedom in a commons brings ruin to all."

Hardin's theory made intuitive sense and provided a temptingly simple explanation for catastrophes of all kinds—traffic jams, dirty public toilets, species extinction. But Elinor Ostrom, a political scientist at Indiana University at Bloomington, was taken aback by its popularity, because her own research showed that freedom in a commons did not necessarily lead to ruin. While Hardin argued that the tragedy of the commons could only be avoided through total privatization or total government control, Ostrom had watched groundwater users near her native Los Angeles develop a means of sharing their coveted resource. Over the next several decades, she studied many other examples of successful community management, including centuries-old systems run by cattle herders in Switzerland, forest dwellers in Japan, and irrigators in Spain and the Philippines.

The common features of these systems, she and her colleagues learned, included clear boundaries (the "community" doing the man-

Elinor Ostrom at work on Manitoulin Island in Ontario, Canada, in 1968, the year Garrett Hardin published his paper "The Tragedy of the Commons."

aging had to be well-defined); reliable monitoring of the water, grass, forest, or other shared resource; a reasonable balance of costs and benefits for participants; a predictable process for the fast and fair resolution of conflicts; an escalating series of punishments for cheaters; and good relationships between the community and other layers of authority, from household heads to international institutions.

When it came to humans and their appetites, Hardin assumed that all was predestined. Ostrom showed that all was possible, but nothing was guaranteed. "We are neither trapped in inexorable tragedies nor free of moral responsibility," she told an audience of fellow political scientists in 1997.

Ostrom's research gained global prominence in 2009, when at age seventy-six she became the first woman to be awarded the Nobel Prize in Economic Sciences. But during the early years of her career, colleagues criticized her for spending too much time studying the differences among systems, and too little looking for a unifying theory. "When someone told you that your work was 'too complex,' that was meant as an insult,'" she recalled. Even social scientists could be impatient with human complexity, and eager for easy answers.

Ostrom insisted that complexity was as important to social science as it was to ecology, and that institutional diversity needed to be protected along with biological diversity. "I still get asked, 'What is *the* way of doing something?' There are many, many ways of doing things that work in different environments," she told an audience in Nepal in 2010. "We have got to get to the point that we can understand complexity, and harness it, and not reject it."

When Ostrom died in 2012, she was celebrated by her colleagues for her pioneering work, her plainspoken humility, and her steady resistance to what she called "panaceas." She knew how corrosive those could be; all her data could not dislodge Hardin's metaphor from the public imagination, and the tragedy of the commons remains a powerful panacea for optimism.

In southern Africa in the late 1980s, a few conservationists began to incorporate Ostrom's ideas into community-based conservation projects. Through the CAMPFIRE initiative, co-founded by University of Zimbabwe researcher Marshall Murphree, district councils earned revenue from safari hunting and tourism on communal lands in Zimbabwe, creating an incentive for the councils to control poaching. In neighboring Zambia, the ADMADE program employed local people as anti-poaching rangers, then transferred some wildlife management responsibilities—and benefits—from the national government to community boards. International aid agencies, along with conservation groups such as the World Wildlife Fund, took notice of these programs' early successes and lent their support.

Owen-Smith, meanwhile, was trying to repair the relationship between the apex species of the desert and their human neighbors.

In the spring of 1983, after a dispiriting first year in Wêreldsend, Owen-Smith met with his old friend Joshua Kangombe, a Herero headman, to discuss the poaching crisis. Kangombe knew that poaching was a problem—for humans and wildlife alike—but he feared it would continue until the deprivations of the drought had eased. "It is easy for us who have full stomachs to talk about protecting wild animals," Kangombe said, "but it is hard for a man to put his firearm away if his children are still hungry." The only solution was for the government to station more conservation officials in the region.

Owen-Smith knew that was unlikely—and even if it were possible, he pointed out, local people were so familiar with the terrain that they could easily evade enforcement. "We both know the government cannot stop the poaching on its own," he told Kangombe. "We need your help." Owen-Smith, who had adopted the local habit of puffing away at a problem, passed his pipe tobacco to Kangombe, and the two sat quietly in the thickening smoke.

Eventually, Owen-Smith asked Kangombe if he would be willing to recommend some reliable local men to act as unarmed game guards. Perhaps, he said, the Namibia Wildlife Trust could supply a regular ration of maize in exchange for their time and their knowledge of the bush. Kangombe, reflecting on the idea, suggested that they not limit themselves to "reliable" men. The local poachers were among the best trackers, and he predicted that they would stop poaching once their children had enough to eat.

When Owen-Smith approached the board of the trust, he proposed that the game guards report not to the trust but to Kangombe himself, as he was already a respected authority. The trustees had their doubts about the plan—even Owen-Smith had reservations—but they agreed to a one-year trial with six men. The men selected were all skilled trackers and well-known in their communities, and Owen-Smith was sure they would be aware of any poaching that took place on the communal lands. He just wasn't sure if they would report it.

Soon after the patrols began, a game guard named Kamasitu Tjipombo reported the poaching of a giraffe in his territory. His sister's grandchild had found a scrap of giraffe skin hidden under a stone, and Tjipombo had followed the poacher's tracks back to where the giraffe had been killed. There, hidden inside a pile of ash, he found a large, charred molar that could only have belonged to a giraffe. Tjipombo also reported that the poacher, a local man, had fled the area. Owen-Smith and two colleagues tracked down the culprit and brought him to court, where he was convicted and sentenced to a year in prison or a heavy fine. At the request of the conservation department official who made the arrest, however, the sentence was conditionally suspended. The point, as Owen-Smith recalled later, was not to put people in jail but to make it clear that poaching would no longer be tolerated.

One of the first game guards recommended by Kangombe was Sakeus Kasaona, a poacher whose tracking skills helped the trust win an early case against a government official who had poached a springbok. Sakeus's son John, who was a boy in the early 1980s, remembers when a white man in a Ford F-150 used to pull up to their house, unload some bags of maize and sugar, and take off with his father in search of elephants to shoot. He also remembers when the Ford was replaced by a well-traveled Land Rover with a lion-track logo on its side doors. The driver of the Land Rover was a white man, too, and he also brought maize, but he was much taller and skinnier, with a bushy beard. And when Owen-Smith and John's father disappeared into the bush, they didn't shoot anything.

The game guards continued to report poaching cases, and as potential poachers learned that residents were watching and reporting their activities, they started to avoid the area. Elephant and rhino poaching in the region stopped completely, and the populations of both species began to recover from the drought and the effects of uncontrolled poaching. Antelope numbers were so improved that Owen-Smith was able to persuade the national conservation department to reopen limited game hunting in the area—a development much appreciated by locals.

Also in the early 1980s, Owen-Smith met Margaret Jacobsohn, a fellow South African who had moved to South West Africa to research her doctoral dissertation in archaeology, and the two became life partners. Where Owen-Smith was immovable, Jacobsohn was unstoppable: impulsive, gregarious, and intensely curious about the desert and its residents. A former investigative journalist, she conducted her research from a dung-plastered hut in the tiny settlement of Purros, where she studied the relationships between Himba people and their material goods. Along the way, she learned just how radically her perceptions of wealth, privacy, and even time differed from those of her Himba acquaintances.

Jacobsohn and Owen-Smith had immediately connected over their shared conviction that successful conservation required local support, but Jacobsohn pushed Owen-Smith to take the game guard program further. While the older men who served as guards were motivated in part by a revived sense of responsibility toward other species, she argued, younger generations would likely need to realize more benefits in order to carry on the work of conservation.

They soon learned that some of the most valued benefits were intangible. In 1987, when Jacobsohn and Owen-Smith attended an IUCN conference on community-based conservation in Zimbabwe, a comment by Harry Chabwela, director of the Zambian national parks, made a lasting impression. "At this conference we have talked a lot about giving local people this and giving them that, but what has been forgotten is that they also want power," Chabwela said. "They want a say over the resources that affect their lives. That is more important than money."

When the conference ended, Owen-Smith and Jacobsohn drove straight to Purros and requested that headman Goliat Kasaona call a meeting of the residents. Owen-Smith started by asking if everyone was satisfied with the occasional tours he had been bringing to Purros as part of his work with a group called the Endangered Wildlife Trust. They were, they said. Then he asked how they felt about the elephants that had recently begun drinking from the Purros spring. "Shoot the elephants, or tell the national conservation department to take them away," people said. The animals were dangerous if threatened, and the competition for water would make it harder for people to grow maize, which they had begun to cultivate after losing most of their cattle to drought.

Like the elderly man who had complained about and then defended "*my* elephants," Owen-Smith suspected that few people really wanted the elephants to be shot. But he understood their frustration with a species they saw as a government responsibility, and felt unable to influence. After some thought, he made a proposal. For every visitor he brought to Purros, the trust would pay the community a fee. The collected fees, as it turned out, were enough to buy maize for everyone, eventually allowing residents to give up their maize gardens and invest more time in their goats and sheep. Jacobsohn also began to work with local women—including several friends from her time in Purros—on crafts that could be sold to visitors. While the cost of living with elephants had not been erased, it had been reduced by material benefits from tourism, and by a growing sense that the animals were not government-owned pests but neighbors whose behavior could be changed.

These experiments—both the game guards and the Purros initiatives—were conducted in what was essentially a war zone. Clashes between South African troops and South West African guerrilla forces had become enmeshed with the civil war over the border in Angola. Local residents, regardless of their own loyalties, were sometimes caught in the middle, forced to shelter soldiers from one side and then punished for doing so by the other.

When Namibia achieved independence in 1990, the leaders of the liberation movement were elected to political office, and many in the government were eager to effect change. Niko Bessinger, the first head of the conservation ministry, was a champion of community-based conservation, and his deputies Brian Jones and Chris Brown recruited Jacobsohn and Owen-Smith to conduct several in-depth surveys of public attitudes toward conservation in the northwestern desert. The surveys, which were supported by the World Wildlife Fund, confirmed what Owen-Smith and Jacobsohn had heard for years. Most people did not want the apex species they lived with to be killed or removed, but, as Harry Chabwela had suggested in Zimbabwe, they did want a say in their management. They knew that white farmers in southern Namibia already held substantial authority over local game and other wildlife, and they wanted similar rights.

Though people in the Kunene Region had often told Owen-Smith that they saw no value in wildlife, it turned out that they saw many values. Like Aldo Leopold, they believed that it was possible to appreciate a species for its own sake and make intelligent use of it, too.

In 1996, the Namibian National Assembly passed a law that allowed groups of people on communal land to establish community conservancies. Conservancies would be governed by elected committees, and members would share the benefits of any tourism or commercial hunting within conservancy boundaries.

Turning law into reality was an enormous and complicated task. Communities, already divided by decades of political and military conflict, were often split over whether or not to participate in the new program. Those that did decide to create conservancies then had to map boundaries, elect leaders, learn basic bookkeeping and negotiation skills, and agree on how to both protect and benefit from the species they lived with. A small nonprofit organization created by Owen-Smith and Jacobsohn, Integrated Rural Development and Nature Conservation or IRDNC, lent technical support to the new conservancies, and founding staffers spent hours facilitating discussions, settling arguments, and training new committee members in unfamiliar procedures. "There was no way through it but meetings and more meetings," remembers Lucky Kasaona, now a regional coordinator for IRDNC.

The first conservancies on communal land were formalized in 1998, and the first joint operating agreements with commercial partners were signed soon afterward. There are now more than eighty communal conservancies in Namibia, stretching from the northwestern desert to the humid, densely populated Zambezi Region in the northeast, and they play a central role in the management of more than forty million acres of land. They earn revenue from lodges, campgrounds, and hunting guide services, both as partners in joint ventures and as solo operators. They participate in annual surveys of game and wildlife populations and, in cooperation with the national conservation ministry, set quotas for both subsistence and commercial hunting within their boundaries. And every year, the members of each conservancy assemble for a general meeting, where they have a chance to call their governing committee to account.

The general meeting of Orupembe Conservancy, held in an open-air pavilion on the dusty outskirts of the tiny town of Onjuva, was supposed to start at 8:30 on a Sunday morning in August, but at two in the afternoon, only a handful of people were seated under the pavilion. Dozens of others were sitting in twos and threes in nearby spots of shade, sharing news.

Onjuva is hundreds of miles from the nearest gas station, and even further from a paved road. Mobile phone service is nonexistent. Most of the people at the meeting were semi-nomadic herders, many of whom had traveled long distances from even more isolated corners of the conservancy. I was present thanks to the expert off-road driving skills of guide Edison Kasupi, who grew up in nearby Purros Conservancy.

When the Onjuva committee finally called the meeting to order, there were ninety-five people seated inside the pavilion, about half of the conservancy members and just enough for a quorum. Chairman Henry Tjambiru commented that the current drought had forced many people to take their herds further afield, preventing them from attending. As Tjambiru began to proceed with business, local headman

Karaere Mupurua stalked to the open space at the front of the pavilion, dragging a low chair across the dirt floor. Though he was bent with age, his eyes were bright, and he cut an imposing figure in his conventional Himba ensemble—a wide leather "hunger belt" that can be cinched up in thin times, a cloth skirt, and car-tire sandals.

"Some of the committee members are not showing respect," he said, pounding his walking stick for emphasis. "We all know who they are, and we're going to get rid of them." A few people giggled nervously as Mupurua plopped down on his chair, crossed his ankles, and stared intently at the committee. Kasupi quietly translated the scolding for me, chuckling as he remembered his own time as a conservancy committee member; he'd endured worse, he said.

Mupurua was referring not only to the late start but to what was forthrightly described on the English-language meeting agenda as "stolen money issues." Two members of the current committee had withdrawn about a thousand U.S. dollars' worth of funds from the conservancy accounts for personal use, and Lucky Kasaona, who was present as a representative of IRDNC, seconded Mupurua by saying that the national conservation ministry was watching Orupembe Conservancy closely. Meekly, Tjambiru thanked Kasaona for the warning. The accused committee members looked defiantly bored.

At first, the assembled conservancy members were quiet, sluggish in the afternoon heat. The crowd inside the pavilion was entirely male except for four young women, cocooned in high-waisted, multilayered Herero dresses, who sat together in an impassive, rainbow-colored row. About a dozen other women, their skin and hair glowing with *otjize*, wore leather loincloths in the minimalist Himba style and sat outside the pavilion on scattered blankets, bouncing infants and talking in low voices.

Orupembe Conservancy has several sources of income, all relatively modest: a campsite, a small lodge that it co-owns with two other conservancies, and contracts with a handful of hunting guides. Another lodge, begun as a joint venture with a foreign investor, has been closed for almost a decade. (Some conservancies have very little income, and fund their operations with donations from international conservation groups; others, like neighboring Marienfluss Conservancy, have joint venture agreements with upscale lodges that can net more than one hundred thousand U.S. dollars a year in salaries and fees.)

After a review of the year's earnings, the committee distributed a list of local species and the current hunting quotas for each. Because the drought had worsened since the quotas were set, conservancy members had voluntarily left most of them unfilled. While wildlife surveys earlier in the year had suggested that seventy-five oryx could be killed without harming the population, for example, only three had been shot so far. The meat from two of those was currently boiling in a nearby row of pots, about to be served for lunch.

Over the next couple of hours, as the sun sank and the day cooled, the pace of business picked up. The committee reported that predators had killed fifty-eight domestic animals over the past year, and that no poaching of any species was known to have occurred. But wait, one member said. Because of the drought, one local family had let its cattle and goats drink from a well designated for wildlife—and because the zebras wouldn't approach the well when the livestock were there, they had been dying of thirst. Shouldn't blocking wildlife from water be considered poaching? This philosophical question sparked a spirited discussion, which was interrupted by an even more spirited discussion about the firing of a local hunting guide by one of the commercial companies operating in the conservancy. The man's friends tried to quiet the gossip, arguing that he should be present to defend himself.

As the disagreement devolved into threats and angry gestures, another local headman stood up and told both sides to leave their "home issues" out of the meeting. Karaere Mupurua, who had been quiet for some time, also stood and asked the young men involved to explain the relevance of the dispute. Getting no clear answer, he threw up his hands. "I'm done!" he cried theatrically. The conservancy members, who were getting restless, took this as a signal to adjourn. Despite protests from the chairman, they began to disperse, some laughing at their unexpected liberation. The stolen money issues would be addressed in the morning—if the meeting reconvened on schedule, that is.

On the surface, it had been a shambles. But as I, too, got up to leave, I realized with surprise that I was exhilarated. Underneath the procedural inefficiency, the petty corruption, and the sideline arguments, real work had been accomplished. Even during an exceptionally difficult year, these conservancy members had taken the trouble to travel to their meeting, consider the long-term future of other species, and recommit themselves to ensuring it. The process might be messy, but it was clearly working in important ways; though the people under the pavilion had lost dozens of domestic animals to drought-starved predators, they had not only stuck to the agreed-upon hunting quotas but chosen to leave some unfilled. I'd attended many meetings with similar goals—and far more material resources—and few could point to such meaningful outcomes. And none could equal the conservancies' decades-long record of successes.

The next day, the conservancy members took a decisive stand on the stolen money, voting out the entire committee and voting in seven new committee members. Among them were the four young women who had brightened the pavilion with their colorful Herero dresses.

Community-based conservation projects, and their results, are as varied as the people who undertake them. In some cases, the community involvement is little more than symbolic, organized by outside groups in order to impress international funders for whom "community conservation" has become fashionable. Even substantive, well-established programs are vulnerable to internal conflict, and to external pressures ranging from drought to war to global market forces. As Elinor Ostrom reminded her audiences, there are no panaceas in conservation, and any strategy can succeed or fail. Community-based conservation is distinctive because modern societies have only begun to understand its potential. "What we have ignored is what citizens can do," Ostrom said.

Ostrom's principles now underlie hundreds of community-based conservation efforts throughout the world. Their members cooperate in the management of marine reserves in the Philippines, highland forests in Cameroon, fisheries in Bangladesh, oyster farms in Brazil, elephants in Cambodia, and wetlands in Madagascar. They operate in thinly populated deserts and crowded river valleys, and they have developed their own rules, many ingeniously suited to their places and circumstances. Some have revived and adapted traditional conservation practices that had fallen into disuse. As the promoters of top-down sustainable development projects discovered in the 1980s and 1990s, true win-win solutions are rare, for conservation usually carries at least some short-term costs. But research by Ostrom's intellectual successors finds that many community-based conservation projects have reduced those costs and, over time, delivered significant benefits, tangible and intangible alike.

During his career in northwestern Namibia, Owen-Smith found, like Ostrom, that well-defined boundaries, reliable monitoring, and predictable consequences are all important to the local protection of species. He often emphasized that everything rested on genuine relationships between people—which, he made clear, should not be confused with alliances between conservation groups and local officials, or with the fleeting connections formed by conservationists who stop by to "train" those with whom they should be collaborating. "You can talk all you want, but if you're not listening, you're not going to get anything done," he told me one afternoon, as we sat drinking strong tea in the narrow shade of his Land Rover. "You have to care about *them* and *their* problems."

There are, to be sure, cases where resentments toward wildlife—or, more frequently, resentments between people about wildlife—are so chronic, or so amplified by social media, that they are extremely difficult to resolve. But most people, regardless of their means and loca-

tion, do not want their local species to disappear forever. Given the
tools to deal with pressing wildlife conflicts, they often come to appreciate once-hated species—elephants, wolves, bison—and will make
considerable sacrifices for their sakes. While conservationists everywhere may still struggle to excite public concern about small, faraway
species like the snail darter, love of other species—and pride in their
protection—is more widespread than they often assume. What Leopold
called the land ethic is often just under the surface, and rises in the
right circumstances.

Bertha Tjipombo, who remembers fidgeting through the early community meetings in Purros as a young girl in the 1980s, is now the
first female leader of her conservancy's game guards. The conservancies, she told me, "help people benefit from what they were losing"—
by compensating those who lose livestock to predators, and sharing
the earnings from tourism and hunting. Like many other conservancy members, though, she said that the most lasting benefits are
immeasurable: "I want my daughter's daughter's daughter to see what
I conserved."

Community-based conservation began as a reaction to top-down
conservation strategies, but it can operate in parallel with large parks
and reserves—and even foster their creation. In Kunene, two neighboring conservancies have proposed to establish a "people's park" where
livestock would be excluded and tourist numbers would be limited by a
permit system, allowing lions and other large predators to more easily
avoid conflicts with humans. Should the national legislature approve
the conservancies' proposal, the region could start to resemble Michael
Soulé's "cores, corridors, and carnivores" vision for the Wildlands Network: a core habitat from which large carnivores can range in relative
safety, since the region's biodiversity is protected not only by law but by
supportive human neighbors.

. . . .

In the early 2010s, South Africa was hit by a massive new wave of rhino poaching. Driven by a spike in Asian demand for powdered rhino horn—a discredited traditional medicine that is still a powerful status symbol—and controlled by international gangs, it caught nearly everyone by surprise. Rhinos were not the only species at risk: because circling vultures can give away a poacher's location, poachers often try to ward off the birds by poisoning the mutilated rhino carcasses they leave behind. The resulting widespread vulture deaths accelerated the already steep decline of the continent's vulture species, whose scavenging plays an essential role in controlling waste and limiting disease transmission between animals and humans.

Because many of the rhino poachers were heavily armed, conservation authorities in South Africa—and in many other African countries—bulked up their arsenals in response, turning park rangers into part-time soldiers. But in Namibia, where the black rhino population was by then second only to South Africa's in size, the conservancies tried a different approach.

Anticipating an onslaught of rhino poaching, Owen-Smith persuaded a few elderly game guards to come out of retirement and mentor the current guards, most of whom were too young to have much firsthand experience with poachers. Save the Rhino Trust, a Namibian organization whose trackers have monitored the Kunene rhino population since the late 1980s, proposed to work with the conservancies to reinforce their game guard system against the new threat. The conservancies agreed, and between August and November 2012 the trust funded the appointment of sixteen conservancy rhino rangers. That December, a rhino was killed within conservancy boundaries—the first rhino-poaching incident in Kunene in decades. Though trackers working for the Save the Rhino Trust quickly caught the poachers and recovered the sawed-off horns, the national police discouraged the trust, the conservancy rhino rangers, and other local residents from participating in future investigations. Over the next two years, twenty-three rhinos were poached in Kunene—10 percent of the region's entire population.

In 2015, Owen-Smith, other IRDNC leaders, and Save the Rhino Trust director Simson Uri-Khob took a group of community elders, local police, conservation ministry representatives, and professional conservationists on a multiday tour of a rhino poaching hotspot. The elders, some of whom had been unaware of the severity of the situation, were alarmed. They held a private, multiday meeting with conservancy committee members, where—reportedly—they reminded their younger listeners of the accomplishments of their forebears and of their continuing responsibility to sustain other species. Committee members, abashed, reenergized their efforts and worked with police, ministry officials, and conservationists to draw up a joint plan. The parties agreed that conservancy rhino rangers, who like the community game guards were unarmed, would carry out patrols with local police. The rangers eventually established an outreach program that encouraged all community members to watch for and report suspected poachers.

The conservancy rhino rangers, whose positions are now jointly funded by the trust and the conservancies and supported by IRDNC, are coordinated by three regional managers, one of whom is Boas Hambo, a nattily dressed Herero man in his late thirties. A former tour guide and IRDNC field officer, Hambo works out of his home office in the small town of Warmquelle, but he is most often on the road, delivering supplies to rangers in the field and talking with people about what they've seen and heard. Because the sixty-six rhino rangers are employed by the conservancies and the conservancies are a trusted institution, locals will often tell Hambo about sightings of unfamiliar cars in the bush or loud-mouthed strangers in the bar, tips that he and the rangers either investigate themselves or pass on to the police. This informal neighborhood watch program works so well that most would-be poachers are warned away from the area before they even get near a rhino. (The resident lions have also proved to be effective deterrents;

since poachers often work at night when lions are active, they're easily spooked by reports of nearby prides.) While hundreds of poachers and rangers have died in the crossfire over rhinos in South Africa and elsewhere, the preventive approach of the conservancy rhino rangers enables them—and the police officers who join their patrols—both to reduce poaching and to avoid risky encounters with poachers.

In 2016, poachers killed three rhinos in Kunene; the following year, they killed four. When I met Hambo in late 2019, it had been a full two years since the last rhino had been poached in the region, but he emphasized that the patrols would continue. "Silence makes me nervous," he told me. "It makes me think the poachers are planning something." His concern was warranted; in April 2020, shortly after Namibia entered a partial lockdown due to the coronavirus pandemic, rangers discovered two rhino carcasses in Purros Conservancy, and two more were discovered in September.

Ricky Beukes, a member of Torra Conservancy, on rhino patrol circa 2014.

I saw only one rhino during my time in Namibia, but I felt lucky. Despite his size, the young male blended into the sparse midday shade of a mopane tree, easy to miss even from two hundred yards away. He worked his droopy upper lip as he stared placidly ahead, seemingly unafraid of our vehicle. Only after several minutes did he turn and walk slowly away, vast haunches jiggling as he disappeared into the sun-bleached brush.

John Kasaona, who as a boy watched Owen-Smith and his father set off on game-guard patrols, is now the executive director of the IRDNC, and while he still spends much of his time in Kunene, he frequently travels beyond it. He has flown to Scotland and Norway to talk with farmers and ranchers about living with predators; he has toured the United States as a guest of the U.S. State Department; he has spoken at the annual TED conference, where he told the story of the Namibian conservancies to an approving audience of the technorati.

What Kasaona mentions briefly, if at all, in his overseas presentations is that the success of the Namibian community conservancy system depends in part on income from trophy hunters—tourists who pay for the privilege of shooting an animal. For many of the conservancies, trophy hunting is not only a source of income but a tool for preserving the peace between humans and other species, since trophy hunters are sometimes directed toward individual lions or elephants who have become aggressive toward people.

Kasaona is well aware of the images that trophy hunting conjures in his listeners' minds: Theodore Roosevelt standing next to a fallen elephant, dwarfed by the carcass and its upturned tusks; Eric Trump grinning as he hefts the limp body of a leopard, his brother Don Jr. beside him; the Zimbabwean lion named Cecil, whose illegal killing by a Minnesota dentist during a guided hunt in 2015 caused a global outcry. For some conservationists in North America and Europe, trophy hunting in Africa has come to symbolize human sins against other species, and even the mention of it can bring the long-simmering conflicts between conservation-inclined sport hunters and other wildlife advocates to a roaring boil.

In 2018, after Kasaona gave a talk about the community conservancies at a Smithsonian Institution conservation conference in Washington, DC, a young woman stood to speak at the audience microphone. "I think that some pieces were missing from the presentation," she began. Kasaona had not shown images of the animals slain by trophy hunters, she said. He had neglected to mention that the lion or elephant spotted by a visiting family on safari might be killed by a hunter the next day.

Kasaona, at the podium, acknowledged the international controversy over trophy hunting, but said that regulated trophy hunts remained an important source of revenue for the Namibian conservancies. "In Namibia, to get to where we are today, is because of sustainable utilization," he concluded. There was more to say, but the session was over, and any further discussion was washed away by chatter.

Almost two years later, I met up with Kasaona in the town of Swakopmund, about halfway down the Namibian coast. Here, a fledgling tourism industry collides with history; on the outskirts of town, picture-windowed vacation homes overlook rows of sandy mounds—the unmarked graves of some of the thousands of people who died in the prison camps run by German colonial authorities in the early 1900s. (The camps, which followed a devastating colonial "war" against the Herero and other groups, are seen by some historians as forerunners of the Nazi concentration camps.)

Kasaona and I talked over dinner at the colonial-era Hansa Hotel, where German is spoken more frequently than English and both are far more common than any of Namibia's twenty-four indigenous languages and dialects. Kasaona ignored the menu, amiably asking the German-speaking server to recommend a dish "for a big Herero man like me." After a few minutes, a Herero server who had overheard the exchange arrived with a thimble-sized dish of chutney, solemnly passing it off as an entrée. Kasaona, after an astonished pause, broke out in laughter and a stream of good-natured Herero insults.

Over generous plates of springbok curry, I asked Kasaona to finish answering his questioner at the Smithsonian conference. "People say, 'I don't like what they're doing to animals,' but most of them wouldn't want to live next to a lion that could harm their family," he said. "People have personalized their relationship with these animals, given them names, but they never talk about how many people were mauled or killed."

The majority of trophy hunters in Namibia pursue more common species such as springbok, whose hunting is permitted through the conservancy quota system. In the case of globally threatened species, the number of animals (if any) that can be shot each year in each country is set by the Convention on International Trade in Endangered Species. In 2004, the parties to the convention approved applications by Namibia and South Africa to allow limited trophy hunting of black rhinos, determining that the population had recovered to the point that five male rhinos could be shot in each country each year. In Namibia, the national conservation ministry chooses which rhinos will be hunted—usually older animals that have become aggressive or territorial—and issues permits for the hunts.

In 2018, a Michigan man named Chris Peyerk paid four hundred thousand dollars to shoot a black rhino in Namibia. When Peyerk, upon his return home, applied for a U.S. Fish and Wildlife Service permit to import the rhino's skin, skull, and horns, public reaction was as scathing as details were scarce. The rhino Peyerk shot lived in Mangetti National Park in northeastern Namibia, which is jointly managed by the national conservation ministry and a local conservancy. The rhino, according to the ministry, was a twenty-nine-year-old male who was blocking younger bulls' access to females. Peyerk's permit fee, which far exceeded what most conservancies earn in an entire year, was deposited into the national Game Products Trust Fund, which offers grants to conservancies and other conservation organizations.

The trophy-hunting system in Namibia isn't perfect, Kasaona acknowledged—there are cases where hunters have killed the wrong

animal—but over the long term, he said, it benefits both the conservancies and the species in question by reducing conflicts between
people and wildlife. When international conservation groups promise to regulate and censure trophy hunting out of existence, Kasaona
hears what he calls "another kind of colonization"—a violation of the
local authority that he and others have spent decades building up, and
a threat to the revenue it depends on. "To say that if Africans choose
utilization without the blessing of the West they will put measures
in place to block us—that doesn't take us anywhere," he said. "What
do they say to the people whose livelihood depends on what they are
trying to ban?"

Global restrictions on trophy hunting, Kasaona argues, are a simplistic response to a complex situation—what Elinor Ostrom might
call a panacea. Not all countries are alike; not all conservancies are
alike; not all conservancy members are alike. Perhaps not even all trophy hunters are, as Rachel Carson once said of Leopold, "completely
brutal." And a few individual lions and elephants are far more dangerous than others, as those who have lost loved ones and livelihoods to
rogue animals can attest.

While those viral images of trophy hunters with carcasses may all
seem to say the same thing, they don't. Some, surely, are symbols of
corruption or needless violence. But in the best cases, they are examples of sustainable utilization: colonial nostalgia, harnessed by the
formerly colonized to further multispecies survival.

Shortly after my trip to Namibia, Owen-Smith learned that the
lymphoma he had lived with for more than two decades had spread to
his liver, and in the early hours of April 11, 2020, he passed away with
Jacobsohn by his side. He was survived by two sons, a grandson, and,
as the IRDNC staff wrote in a statement immediately after his death,
"the many Namibian conservation field workers and leaders who were
shaped and mentored by Garth over many decades." Owen-Smith had

Garth Owen-Smith and Margaret Jacobsohn in Omatendeka Conservancy, Kunene Region, January 2020.

not only spent half a century advancing local conservation leadership in Namibia, but had made sure it would outlive him.

Namibia's conservancies can't solve climate change; they can't solve poverty; they can't reduce the global demand for illegal wildlife products, or control the violent syndicates that dominate the trade. They can't protect their members from shortsighted industrial development or from economic shocks such as the coronavirus pandemic, which obliterated the conservancies' tourism earnings in 2020. They have, however, restored an essential layer of authority to the conservation pyramid, and the political power they return to communities can be exercised against larger threats. When the Namibian government, in cooperation with foreign investors, proposed to dam the region's largest river in the early 1990s and again in the late 2000s, organized local opposition, centered in the conservancies, helped check the projects.

The real limitation of community-based conservation, in Namibia and elsewhere, is that the relationships and institutions it requires take years—sometimes generations—to develop, and many species can no longer wait for rescue. Emergency measures can head off extinction, but they seldom restore species to abundance. For conservationists at work today, there is new urgency in the maxim of Willard Van Name and Rosalie Edge: The time to protect a species is while it is still common.

Chapter Nine

THE FEW WHO SAVE
THE MANY

Toms Creek, which skirts the suburban edge of Blacksburg, Virginia, is in stretches slim enough to step across and barely deep enough for wading. Standing on the grassy bank, inside a shady green tunnel of trees, I felt a little silly as I raised a pair of binoculars to my face, training them on the clear water a few feet away. But as the view came into focus, what I saw was so beautiful and strange that I caught my breath.

Below the surface of the water, dozens of thumb-sized black-and-yellow fish were massed against the current, wriggling furiously in formation. Together, they hovered over a pile of pebbles, jostling one another as they fought to stay in place. Every few moments, a larger, paler fish approached, a pebble bigger than his head gripped in his jaws. As he gently added the pebble to the pile, the black-and-yellow fish kept wriggling. The whole scene was no larger than a Frisbee, but it had the busy mystery of a tropical reef, filling my field of vision with color and motion.

Emmanuel Frimpong, standing beside me in hip-high rubber waders, grinned at my surprise. In the British-inflected English of his native Ghana, he narrated the scene. The pebble-carrying fish were bluehead chub, and the pebble piles were their nests, which they build each spring and which can sprawl for yards across the creekbed. The nests are used not only by the chub but by nine other species of fish, at least seven of which also lay their eggs in the pebbles and take part in protecting the nest. The black-and-yellow fish were mountain redbelly dace, and their frantic dance keeps silt from settling on the nest and suffocating the eggs; fish from other species lingered nearby, flicking away silt or guarding against crayfish and snapping turtles. A single nest can be surrounded by hundreds of fish—a motley school ranging from pale blue and pink to brilliant yellow and red.

Surrounded by mountain redbelly dace, central stonerollers, and white shiners, a bluehead chub (top) prepares to place a rock on its nest in Toms Creek.

Frimpong has been observing these nests for a decade, and says he could keep learning about them for decades more. But he came upon them almost entirely by accident. When he joined the Fish and Wildlife Conservation faculty at nearby Virginia Tech, he began studying the bluehead chub mostly because it was both abundant and close to campus—key advantages for a beginning researcher on a budget. During their initial surveys of the creek, Frimpong and his students saw the piles of pebbles in the streambed, and gradually realized that they were the base camps of an intricate campaign for survival. The common fish of Toms Creek were engaging in some very uncommon behavior, occasionally observed elsewhere but rarely studied closely.

Toms Creek flows through a town park that is popular with joggers, dog-walkers, and birdwatchers, and passersby sometimes stop to talk with Frimpong and his students. Though the chub nests are visible from several spots along the park paths, visitors are invariably surprised to learn what's happening in the creek. I asked Frimpong to describe their most frequent reaction, and he smiled.

"Joy," he said.

Love of other species often begins with childhood experiences of plenty—fish flashing in a neighborhood creek, herds of elk or wildebeest migrating across a familiar valley, the flocks of birds that young Aldo Leopold watched on his native Iowa prairie. Many early conservationists sought to sustain that abundance. The women who fought the plume trade wanted to save the rich avifauna of the southeastern United States. Rosalie Edge established the Hawk Mountain Sanctuary to protect then-common raptor species during migration. Conservation-minded hunters lobbied for game laws that would benefit their sport and, in many cases, preserve healthy populations of the animals they pursued.

But as conservationists became more alert to global extinctions, the IUCN Red List and, later, the U.S. Endangered Species Act and similar laws drew attention away from sustaining abundance and toward preventing extinction. By the time Michael Soulé founded the "crisis discipline" of conservation biology in the 1980s, conservationists were indeed motivated by crisis, and the crisis in question was an expected onslaught of human-caused global extinctions.

We know that over the last five hundred years, the planet has lost 755 animal species and 123 plant species. The list includes the dodo, the passenger pigeon, the Carolina parakeet, the Pinta giant tortoise of the Galápagos Islands, and lesser-known species such as the longjaw cisco, a foot-long fish from the Great Lakes that was one of the first species to be declared extinct under the Endangered Species Act. Thousands of other species linger in the "extinct in the wild" and "critically endangered" categories of the Red List, including nearly three hundred of the amphibian species decimated by the global pandemic of fungal disease that emerged in the 1980s.

We also know that this tally of extinctions and near-extinctions is incomplete. Despite the diligent work of Linnaeus, his disciples, and subsequent generations of taxonomists, most of the estimated nine million species on earth have yet to be formally described and named, and fewer than one hundred thousand have been thoroughly surveyed. Species that scientists have never documented, or seldom watch, are quietly being obliterated by human activities; some, like the least vermilion flycatcher of the Galápagos, have been recognized only after the death of their last living representative.

Estimates of how many species are going extinct, and how quickly, are complicated by multiple unknowns. In the 1990s, ecologist Stuart Pimm and colleagues calculated the extinction rate to be between one hundred and one thousand times the pre-human "background" rate, which they conservatively estimated to be one extinction per million species per year. More recent studies of the fossil record have concluded that the background extinction rate is likely much lower, pushing the current rate of human-caused extinctions toward the upper end of Pimm's estimate: one thousand extinctions per million species per year. Given our best guess at the number of species on earth, that's nine thousand human-caused species extinctions every year.

Paul Ehrlich is one of many biologists who say we are close to or already entering the sixth extinction—a mass extinction that, unlike the five mass extinctions the planet has already experienced, is primarily caused by humans. As a biologist who focuses not on species but on the populations that compose them, Ehrlich is particularly aware that the current count of extinct species, while sobering, doesn't reflect the localized declines and extinctions that are already affecting ecosystems and foretell worse to come. "Species extinctions are difficult to find, but population extinctions are easy—they're happening before our eyes." he told me. "The extinction of a species is the last stage of a process that has already robbed you of much of what you want."

In a 2017 analysis of nearly twenty-eight thousand vertebrate species—almost half of all known vertebrates—Ehrlich and his colleagues Gerardo Ceballos and Rodolfo Dirzo showed that a third had decreased in both population size and geographic range since the year 1900. When they looked more closely at 177 mammal species, they found that all had lost at least 30 percent of their range over the same time period, and more than 40 percent had experienced severe population declines. Writing in the *Proceedings of the National Academy of Sciences*, Ehrlich and his co-authors characterized the trends as a "biological annihilation": "Our data indicate that beyond global species extinctions Earth is experiencing a huge episode of population declines and extirpations, which will have negative cascading consequences on ecosystem functioning and services vital to sustaining civilization."

The leading causes of this annihilation haven't changed much since William Hornaday was raising bison in the Bronx: people are still killing too many animals and destroying too much habitat. A 2016 study of more than eight thousand threatened and near-threatened animal and plant species on the Red List found that the most common danger faced was overexploitation—illegal or unsustainable hunting, fishing, logging, and plant gathering—followed closely by the destruction of habitat for food, fodder, or fuel.

These long-standing threats are now amplified by the effects of climate change. Rising global temperatures are shifting species ranges toward the poles, both on land and in the sea, and while some species can adapt, others cannot. Heat and drought are worsening forest pest infestations and leading to more destructive wildfires. Coral reefs, which provide food and shelter for one in every four marine species, are threatened by ocean acidification—caused by the absorption of carbon dioxide from the atmosphere—and by rising water temperatures; in 2016 and 2017, the marine park that encompasses the Great Barrier Reef could not defend it from back-to-back heat waves that severely damaged half its coral. And as heat, floods, and droughts make it harder for our own species to grow food and find shelter, we are likely to put more pressure on others, killing yet more animals and destroying yet more habitat. In 2019, a global assessment by an international panel of biodiversity experts estimated that a million species were at risk of going extinct within decades—including as many as a quarter of all animal and plant species.

The concept of conservation biology as a crisis discipline, and of conservation as a crisis response, has undoubtedly contributed to the protection of both populations and species. While conservationists are notoriously bad at measuring their own progress—there is, after all, always another crisis to attend to—one analysis found that since the 1980s, conservation efforts have slowed vertebrate species' tumble toward extinction by at least 20 percent overall. And conservation efforts over the past century and a half have, of course, averted the extinction of many individual species, from bison to bald eagles to black rhinos. The IUCN is developing a "Green List" of recovered species in hopes of calling more attention to these successes, and inspiring others.

But crises call for emergency measures, and emergency measures are almost always both costly and temporary—barricades against disaster, enacted with little consideration of root causes or long-term consequences. The larger project of conservation—that of protecting the relationships that support all life on earth—can't be accomplished with emergency measures alone. It has to start with common species.

Emmanuel Frimpong grew up in southern Ghana, in what was once the heart of the powerful Ashanti Empire. When he was a boy in the 1980s, his hometown of Obuasi was dominated by one of the largest gold mines in the world. Frimpong's parents were teachers, but more than half the city's adult population worked directly for the mine. When he heard the phrase "natural resources," he thought not of animals or plants but of gold.

When Frimpong was about ten, his father and uncle began letting him tag along on their fishing outings—ten-mile hikes to a stream that hadn't yet been polluted by the mine. They fished with hooks, some twine, and long sticks, and every fish they caught was food. Soon, Frimpong and his friends started to make the trek on their own, catching small fish with homemade poles and live-trapping common birds to sell in the market as pets.

Like many middle-class Ghanaians, Frimpong's parents hoped their son would become a doctor or an engineer ("a career catching birds was not exciting to them," he remembers dryly), but admission to medical school was tough, and after one failed attempt he decided to enroll in the newly established renewable natural resources program at Ghana's main science and technology university. While the program focused on managing fisheries for food and forests for timber, he learned that other species also needed protection. During a stint as a research assistant for one of his professors, he spent three months identifying all the animal parts for sale in the local market, counting hundreds of deer tails, mongoose paws, and porcupine quills. He encountered the remains of pangolins and other species that he knew were in decline.

Ghana, tucked under the chin of Africa on the Gulf of Guinea, declared its independence from Britain in 1957, kicking off the continent-wide independence movement that so concerned Julian Huxley and his fellow European conservationists. Though both the colonial and national governments enacted forest protection laws and regulations, Ghana is estimated to have lost nearly ten million acres of forest to logging and agricultural development during the twentieth century. Many animal species that once occurred throughout the country are restricted to parks and reserves, and even there, subsistence hunting has depleted mammal populations.

So, in 1999, when Frimpong traveled to Kenya for a month-long tropical biology course, he was as astounded by the variety of its wildlife as any other foreign tourist. He was aware that Kenya's national parks rarely benefited local people, but he saw that they had tremendous potential to support both economies and ecosystems. Protecting threatened species, he realized, could be more than a passion—it could be of practical help to people, too.

Since there were no graduate programs in conservation biology in Ghana, Frimpong started looking abroad. In the early 2000s, he and his wife moved to the United States, where he did graduate work on the environmental impacts of aquaculture and, later, the effects of land use on freshwater fish communities. From the start, he was as interested in the social and economic aspects of conservation as he was in the biology of the species he studied, a breadth that gave him some difficulty when he entered the academic job market. "People said, 'Are you a fisheries biologist or a fisheries economist?'" he recalls.

When he arrived at Virginia Tech, he understood that he needed to, as he says, carve out his own ecological niche—secure funding for research that would simultaneously advance his career, attract graduate students, and contribute to conservation. The established fisheries biologists in his department were already studying game species, endangered species, marine species, and larger river species. One of the few unoccupied niches was in Toms Creek, among the bluehead chub—where he found not only a research project but a new perspective on conservation. "School taught me about endangered species," he says. "The fish educated me about the importance of common species."

Emmanuel Frimpong at work in Toms Creek, June 2020.

Ecologists who study relationships among species have traditionally focused on competition and predation—because of their established importance to evolution, and because they are often more obvious and easier to document than cooperation. Researchers have paid relatively less attention to mutually beneficial relationships, called mutualisms, and reports of them have sometimes been met with the same kind of skepticism that greeted Elinor Ostrom's work on human cooperation. On and around the bluehead chub nests of Toms Creek, however, Frimpong and his students found that while fish of different species often challenge one another on first encounter, they quickly settle into a détente, joining a collective that serves at least ten of the creek's fish species.

The value of the bluehead chub can't be measured in scarcity, or calories, or tourism potential. We can appreciate its cooperative instincts, wonder at the small splendor of its nests, and assume that it contributes to the various services provided to us by its ecosystem, but its role in our own lives is to a large extent "inscrutable to us," as Rachel Carson once wrote of the "wisp of protoplasm" known as the sea lace. For the fish who live alongside the bluehead chub, however, this unprotected, unremarked species is indispensable.

The chub reminds me of a story that Garth Owen-Smith liked to tell about his early years in the Namibian desert. One day, while he and a colleague were surveying the medicinal and other practical values of local plants, they asked an elderly Himba man about the uses of a common tree. Through a translator, he replied that he didn't use that kind of tree for anything. So it has no value, Owen-Smith's colleague remarked. The man protested that he had said nothing of the kind. Of course it has value, he scolded. The birds use its branches to sit on.

Upstream of the first nest that we spotted in Toms Creek, the water was deeper and colder, and it gripped my hip waders as we picked our way along the channel. The nests were still clearly visible on the silty bottom; some were small and quiet, abandoned after a recent cold snap, while others were piled with tens of thousands of pebbles and swarming with fish.

Since Frimpong began studying the bluehead chub, he's pulled together reports of mutualistic nest-building and breeding behavior in freshwater fishes on nearly every continent. He's certain that he'd find it

in Ghana, too, for as a boy he often caught small fish with head tubercu-
les resembling those of the bluehead chub. But in tropical Africa, where
the streams are turbid with silt, fish surveys are more complicated than
a stroll along the bank. Very few of the region's smaller freshwater fish
species have been described in detail, and fewer still have been assessed
for the Red List. Any one of them might be crucial to the survival of other
fish species, including those that humans depend on for food. And in
Ghana, where fish are a primary source of animal protein, healthy fish
populations relieve hunting pressure on mammals.

Frimpong has already established research projects in Ghana and
Cameroon, and in the coming years he hopes to spend more time in
Africa. He and his wife have maintained ties in Ghana through regular
visits, and his two children, who have grown up in Blacksburg, have
a strong connection to their parents' home country. Frimpong wants
to develop a more comprehensive fish research program in tropical
Africa, one that will seek to understand the web of relationships within
the cloudy waters of its freshwater streams. "If I didn't do that," he told
me, "I'd feel that I hadn't done something I set out to do."

Over the past century and a half, the conservation movement has
managed to provide some sanctuary for game species (through hunting regulations and wildlife refuges), for many bird species (through
national laws and international treaties), and for some threatened and
endangered species (ditto). What it hasn't figured out how to do, in
any systematic way, is protect everything else. "Nature advocates have
obtained much of what they have asked for," writes legal scholar Holly
Doremus, "but they have not asked for what they really want."

The existing protections for common species are, for the most part,
a side benefit of protections for declining or edible species. Safeguards
designed for an endangered apex species can also protect a suite of
common species. Wildlife refuges established for game species, or
endangered species, can shelter common species, too.

The Ramsar Convention on Wetlands of International Importance, signed in 1971 at the Iranian seaside resort of Ramsar, was motivated by a widespread concern for waterfowl, but it provides for the protection of habitats ranging from peat bogs to mangrove swamps to coral reefs—places rich in both threatened and common species. The Bern Convention, which came into force in 1982 and has been ratified by nearly all European nations and four African countries, emphasizes protections for threatened species but affirms the importance of all species, requiring—not recommending—that its parties "maintain the population of wild flora and fauna."

Parties to the 1992 Convention on Biological Diversity, the most significant international biodiversity protection agreement, also acknowledged the importance of all species and the value of an "ecosystem approach" to conservation. Ratified by all of the United Nations member countries except the United States, ten-year conservation targets set under the convention in 2010 led to a significant expansion of protected areas, particularly marine reserves. Like most international agreements, however, the success of the Convention on Biological Diversity depends on the goodwill of its parties. The new reserves have few concrete benefits for biodiversity, researchers have found, and the parties failed to meet their targets.

Like the U.S. Endangered Species Act, these and other international conservation initiatives and agreements are essential but not sufficient. The reality is that our current local, national, and international protections for common species don't do enough to keep them common. As a result, they don't do enough to keep humanity within its planetary boundaries, or stave off the sixth extinction. One recent study found that while North America's birds of prey and waterfowl have benefited from sustained conservation attention over the past century, the continent's overall number of birds has declined by nearly a third since 1970—a loss of some three billion birds. "The overwhelming focus on species extinctions," the authors wrote, "has underestimated the extent and consequences of biotic change."

Protections for common species could certainly be strengthened with more funding, broader authority, or more popular support, and conservationists have worked for decades to defend and incrementally expand them. But how can conservationists obtain what they really want— which is to conserve species in abundance, in place, and in perpetuity? One way is to rebuild local conservation institutions, as the community conservationists of Namibia have done, and through them reawaken humans' sense of responsibility toward all species. Another is to work within existing institutions—and use them to tell a different story.

In October 1971, a few months after international officials gathered in Iran to sign the Ramsar wetlands treaty, Christopher Stone was standing in front of his introductory property law class at the University of Southern California's law school, trying to hold the attention of his restive students. Observing that the definition of property had changed radically over human history, transforming not only the distribution of power within society but society's view of itself, he wondered aloud about the effects of a similarly radical redefinition of "rights." What if legal rights were extended to, say, rivers? Or animals? Or *trees*? "This little thought experiment," Stone recalled decades later, "was greeted, quite sincerely, with uproar."

Class was soon dismissed—to the relief of the students and their professor—but Stone did not abandon his thought experiment. Instead, he called the law library's reference desk and asked if there were any pending cases where the "rights" of a natural object might affect the outcome. Within half an hour, a librarian called back to suggest *Sierra Club v. Hickel*.

The case involved a proposed Walt Disney resort in Mineral King Valley, a wilderness area in the Sierra Nevada of California. The Sierra Club Legal Defense Fund had challenged the development permit that the U.S. Forest Service had granted to the Disney corporation, but the court that heard the case had ruled that the Sierra Club had no "standing"—because the club itself would not be "aggrieved" or "adversely affected" by the development, it had no right to sue. The club had appealed, and the case would soon reach the U.S. Supreme Court.

Stone hastily pulled together an essay for an upcoming issue of the *Southern California Law Review*. "I am quite seriously proposing that we give legal rights to forests, oceans, rivers, and other so-called 'natural objects' in the environment—indeed, to the natural environment as a whole," he wrote. He was not, he emphasized, suggesting that no one be allowed to cut down a tree. "To say that the natural environment should have rights is not to say that it should have every right we can imagine, or even the same body of rights as human beings have," he wrote. "Nor is it to say that everything in the environment should have the same rights as every other thing in the environment."

Stone's argument was that natural objects should not have to depend on the Sierra Club for legal standing, but should be accorded their own. If forests, oceans, and rivers possessed standing, Stone argued, court-approved human "guardians" could sue on their behalf, much as legal representatives can sue on behalf of incapacitated people or defunct corporations. These guardians would represent the interests of the natural objects, and the court would award any compensation to the object itself, perhaps in the form of funds for restoration.

Like the rights held by people—such as the right to vote and the right to own property—many of the existing protections afforded to natural objects are open to interpretation, and can be limited in certain circumstances. Legal standing for natural objects, Stone predicted, would allow their human advocates to spell out the consequences of such limitations, and elevate those consequences in the eyes of the courts and the public.

"Should Trees Have Standing?" was published, by design, in an issue of the law review for which Supreme Court Justice William O. Douglas had agreed to write a preface, and it found its way to its intended audience of one. Though the Supreme Court ruled against the Sierra Club in April 1972, Douglas cited Stone's argument in the opening lines of his dissent. "Contemporary public concern for protecting nature's ecological equilibrium should lead to the conferral of standing upon environmental objects to sue for their own preservation," Douglas wrote. He concluded his dissent by quoting Leopold: "The land ethic simply enlarges the boundaries of the community to include soils, waters, plants, and animals, or collectively: the land." (Public opposition to the Disney development continued after the court ruling, and company executives eventually lost interest in the project. The Mineral King Valley, now part of Sequoia National Park, remains undeveloped.)

Douglas's unusual dissent caught the attention of the press, and Stone's argument was widely ridiculed. "If Justice Douglas has his way— / O come not that dreadful day— / We'll be sued by lakes and hills / Seeking a redress of ills," one lawyer wrote. A Michigan appeals court later issued an entire opinion in doggerel, beginning with the lines "We thought that we would never see / A suit to compensate a tree."

Stone's essay continues to be argued over by law students, but fifty years after its publication, the notion of "a suit to compensate a tree" no longer seems so far-fetched. The U.S. Endangered Species Act, signed into law two years after Stone published his article, includes a provision that allows private citizens to sue over alleged violations of the act, and courts have since granted standing to spotted owls, red squirrels, Florida Key deer, and other endangered species as co-plaintiffs in citizen suits.

Charles Darwin, in *The Descent of Man*, observed that, over the course of human history, the "sympathies" of *Homo sapiens* have become "more tender and widely diffused, so as to extend to the men of all races, to the imbecile, the maimed, and other useless members of society, and finally to the lower animals." History, of course, doesn't move in a straight line, and it's clear from our own time that sympathies can contract as quickly as they expand; in 2020, the U.S. Fish and Wildlife Service proposed to severely curtail the protections granted to birds more than a century earlier by the Migratory Bird Treaty Act. Yet for many groups of humans—women, children, people of color, people with disabilities, people without property—rights now widely considered inalienable were not so long ago seen as ridiculously out of reach. And our sympathies for other species are already being translated into legal rights.

In 2017, the Whanganui River—the longest navigable river in New Zealand—became a legal person. The Whanganui had been controlled by colonial and national governments since the mid-1800s, but the indigenous Māori people had never willingly given up their historical claim to the river, and it had become the subject of one of the country's longest-running court battles. After Māori legal scholar Jacinta Ruru and her student James Morris noted in a 2010 article that Stone's concept of legal standing for natural objects resembled the Māori concept of rivers as living beings, government officials and Māori leaders took up the idea: if they couldn't resolve who owned the river, maybe the river should, in a sense, own itself. In 2014, the parties reached a settlement that recognized the Whanganui, its tributaries, and "all its physical and metaphysical elements" as "a living and indivisible whole." Three years later, following a parliamentary debate conducted in Māori and English, the settlement became law.

The Whanganui, which has been dammed for hydropower, mined for gravel, dynamited for steamer passage, and polluted by urban effluent, will now be represented in court by two guardians, one chosen by the Māori groups most closely connected to the river and one by the national government—not unlike the arrangement Stone proposed. Exactly what legal personhood means for the Whanganui will surely be the subject of lawsuits to come. But the agreement changes the terms of those arguments. "The law, like our most fundamental societal stories, reminds us not only of what we are, but of what we aspire to be," writes Holly Doremus. "Stories that become embedded in law are thus powerful forces in shaping society and social attitudes." Often, they reinforce existing ideas. Sometimes, they expand our sense of the possible.

Though the Whanganui legislation and a growing number of laws like it vary in their particulars, they replace what Stone called "the view that Nature is a collection of useful senseless objects" with the general proposition that rivers, or forests, or animal species have value beyond the services they provide to humans, and that the responsibilities and costs of tolerating and sustaining them should be borne not by an unlucky few but by society as a whole.

New Zealand biologist and anthropologist Mere Roberts, herself of Māori descent, writes that the Māori worldview might be best described as "kin-centric," in which humans are bound to one another, and to the rest of the living world, by a multitude of reciprocal relationships. Like Leopold's concept of plain citizenship, this view recognizes that some of our relationships with other species are consumptive, some are predatory, and some inflict costs on both parties. It also recognizes that all have at least some potential for reciprocity, if only in the broadest sense of continuing the flow of energy within the biotic pyramid.

As individuals and as species, living organisms are part of interdependent communities, existing within a web of mutualisms that Leopold once imagined as "a universal symbiosis." Given the harm our species is capable of doing to others, it's understandable that over the course of the conservation movement, some have tried to sever our relationships with other species, drawing hard boundaries in an attempt to limit our exploitation of other forms of life.

Boundaries have been useful to conservation—and will continue to be. But the lesson of ecology, much like that of Aesop's fables, is that human relationships with the rest of life are both inescapable and inescapably complex. The great challenge of conservation is to sustain complexity, in its many forms, and by doing so protect the possibility of a future for all life on earth. And for that, there are no panaceas.

HOMO AMPHIBIUS

On a sunny fall day in southern California, deep within the San Diego Zoo Safari Park, a two-and-a-half-month-old baby rhinoceros, Edward, was gamboling around his mother, Victoria. When Victoria gave birth to Edward, she weighed about five thousand pounds and he weighed one hundred and forty-eight. Her weight hadn't changed much in the weeks since, but his had tripled and then some—to five hundred pounds—and two tiny nubs of horn were rising on his snout. When Victoria approached the gate of their fenced enclosure, where a bucket of hay was on offer, Edward bustled over, too. He poked his huge, dusty jaw through the steel bars, snorting as he submitted to affectionate scratches from human hands.

Edward is a southern white rhino, a subspecies related to the southwestern black rhinos protected by the community conservancies of Namibia. While both subspecies are threatened by poaching, neither is in immediate danger of extinction. Edward, though, is a singular animal—the first white rhino born via artificial insemination in North America—and he represents a step toward the still-distant goal of recreating a free-roaming population of northern white rhinos, a subspecies that suffered decades of uncontrolled hunting and poaching and is now functionally extinct.

Researchers at the zoo and elsewhere hope to "rewind" the extinction of the northern white rhino by using genetic techniques to turn frozen skin cells into stem cells and then into viable sperm and eggs. If they are successful in creating northern white rhino embryos, they will implant them in the uteruses of surrogate southern white rhino mothers, raise the resulting northern white rhinos in captivity, and eventually—assuming the risk of poaching can be reduced—reintroduce northern white rhinos to their native range in central Africa.

But none of these high-technology husbandry techniques has been fully developed for northern white rhinos—or any rhinos, for that matter. The genetics research remains in its early stages. The standard method of in vitro fertilization requires too large a fraction of the tiny existing reserve of frozen northern white rhino sperm, so an alternative method has been adopted. Embryo implantation is complicated by the rhino cervix, whose canal is so tight and twisted that reproductive scientists describe it as "tortuous." The zoo has enlisted a team of engineers to develop a robotic catheter that can reach inside a female rhino's two-foot-long vagina, wind through her cervix, and place an embryo inside her uterus. The device looks like an oversized dental saliva ejector and writhes like a snake.

In early 2018, technicians successfully implanted a southern white rhino embryo into Victoria's uterus, and after a sixteen-month pregnancy and a thirty-minute labor, she gave birth to Edward on July 28, 2019, demonstrating that the zoo's fertilization and implantation techniques could be used to produce a healthy baby rhino. Edward's birth made national news—including a mention in the "Pets" section of *People*—and in San Diego it was celebrated by zoo scientists, the rhinos' team of devoted human keepers, and much of the city.

The zoo has a history of what Oliver Ryder, its director of conservation genetics, calls "hopeful interventions," most famously playing a central role in the captive breeding and release of the California condor in the early 1990s. As excited and pleased as Ryder and his colleagues have been by Edward's birth and subsequent advances in their research,

Three-week-old Edward leaps through the air at the San Diego Zoo Safari Park's Nikita Kahn Rhino Rescue Center, August 2019.

they all know that years, perhaps decades, of costly work remain—and that failure is far more likely than success. They also know that success in the laboratory and the rhino enclosure will only take them to the starting line of traditional species conservation. All the complexities of controlling poaching and protecting habitat—of providing a new population with enough food, water, space, and security to reproduce, adapt, and persist—will still lie ahead.

To those of us outside the laboratory, far from the laborious cycle of trial and error required to develop any new genetic or reproductive technique, it's tempting to invest too much hope in these efforts. Wouldn't it be wonderful if we could bypass the gradual, uncertain, politically fraught struggle to protect species and their habitats and simply . . . make more animals? Those who know better say it's hardly simple. Mark Vermeij, a marine biologist on the Caribbean island of Curaçao, is part of an international effort to raise endangered coral larvae in captivity and use them in the restoration of degraded reefs. "People have approached us and said, 'Ah, that's nice, because now the Great Barrier

Reef is fine,'" he told me. "And it's like, 'What on *earth* are you f—ing talking about?'"

During their lifelong conversation, the brothers Julian and Aldous Huxley often described human beings as amphibians. For Julian, the metaphor expressed his hope that humans would soon take charge of their own evolution, transforming themselves into a truly special species. For Aldous, it described the enduring twofold nature of humanity: "Whether we like it or not, we are amphibians, living simultaneously in the world of experience and the world of notions, in the world of direct apprehension of Nature, God, and ourselves, and the world of abstract, verbalized knowledge," he wrote. "Our business as human beings is to make the best of both these worlds."

For the past century and a half, the conservation movement has worked in the world of experience and the world of notions, drawing from both to provide sanctuary to other species. While the morals of its stories will evolve, as the morals of Aesop's fables have, conservationists have established the importance of protecting not only other species but the relationships among them, of not only preventing extinction but protecting possibility. From ecology and conservation biology, conservationists have learned what other species need to persist; from the social sciences, and from real-world experiments like those undertaken by the community conservancies of Namibia, they are learning how humans can more equitably distribute the burdens and benefits of meeting those needs.

Now, though, the ground is shifting beneath us all. In the mid-nineteenth century, Europeans and North Americans grappled with the news that while their own species was not as special as they had once believed it to be, it was powerful enough to pursue other species to extinction. In our century, the unwelcome tidings are that humans can also upend the global climate and acidify the oceans, enabling us to disrupt ecological relationships and destroy species not only one by one but on a mass scale. At the same time, new gene-editing tools are magnifying our influence over other species, allowing us to dream of rewinding extinction and enabling even kitchen-table chemists and middle school science students to synthesize new forms of life. Given our burgeoning ability to destroy and create, we might be forgiven for wondering if we are, Darwin notwithstanding, divine.

Biochemist Jennifer Doudna, one of the pioneers of the CRISPR gene-editing technique, summed up the implications of synthetic biology in 2017: "We're standing on the cusp of a new era, one in which we will have primary authority over life's genetic makeup and all its vibrant and varied outputs." Julian Huxley would be delighted, and by any measure, the prospects are tantalizing: with the tools of synthetic biology, enthusiasts speculate, humans might eliminate hereditary disease, bring about a new and improved Green Revolution with low-input, high-yielding crop varieties, and replace fossil fuels with carbon-free, bioengineered energy sources. Other species could bask in the benefits of our genius.

Faced with these possibilities, it's tempting to set aside the maxims of conservation. Why settle for Aldo Leopold's vision of plain citizenship when some of us might rule benevolently? Why stay mired in complexity when the answers seem so clear? We live in a time of crises, after all, and crises call for swift action on grand scales. As environmentalist and entrepreneur Stewart Brand famously observed in 1968, "We are as gods and might as well get used to it."

The problem, as Doudna herself acknowledges and the rest of us face daily, is that *Homo sapiens* is monumentally unprepared for divinity. As the presumptive drivers of evolution, we may not be dumb, but we are often drunk. We are easily distracted, and chronically shortsighted. We are clever cousins of the mushrooms, promoted beyond our experience.

Leopold, in an essay he wrote after "The Land Ethic," compared Wisconsin—and, by extension, the world—to the mythical Round River, where woodsman Paul Bunyan undertook his eternal odyssey. "We of the genus *Homo* ride the logs that float down the Round River, and by a little judicious 'burling' we have learned to guide their direction and speed," he wrote. Yet we are still carried by the current, he cautioned, and our navigation is shaky: "We burl our logs of state with more energy than skill."

One response to the prospect of technological divinity is to turn away from it, to insist that humans—or some humans—are just another species. We need our distinctiveness, though. We need to pursue technologies that can help us feed ourselves, stabilize the climate, shrink the collective human footprint, and better protect more species from our own. We're capable of restoration as well as destruction, reasoned decisions as well as thoughtless consumption. To insist otherwise, to deny the full range of our complexity, would mean abandoning responsibility for the damage already done—and giving up on the hard-won potential of conservation.

Another response is to see our species as Aldous Huxley did, as amphibians eternally hopping between experience and notions, matter and abstraction. Our dependence on the rest of life can remind us of our vulnerability, and help us use our influence with humility. There is no sinking back into the safety of the pond—humans have already altered "nature" in countless ways, and will continue to do so—but we can approach uncharted technical and ethical territory with the caution of the rhino geneticists in San Diego, alert to past mistakes and present uncertainties. We can engage in high-technology husbandry without mistaking it for what conservationists really want.

In a foreword to the 1946 edition of *Brave New World*, Huxley looked back at the novel he had published fourteen years earlier. "Remorse," he wrote, "is as undesirable in relation to our bad art as it is in relation to our bad behaviour." He would, he promised, limit himself to mentioning what he saw as "the most serious defect in the story": the fact that its main character is forced to choose between a hyper-engineered "bad utopia" and a brutal pre-technology society.

Huxley wished he had offered his character a third alternative.

"Between the utopian and the primitive horns of his dilemma," he wrote, "would lie the possibility of sanity." Writing in the aftermath of the Second World War, Huxley imagined a sane society as one "composed of freely co-operating individuals," in which humans used science and technology, not the other way around.

Novelist Margaret Atwood, in a reflection on the modern relevance of *Brave New World*, remarked that "Alone among the animals, we suffer from the future perfect tense." As both Atwood and Aldous Huxley have so memorably done, we can imagine the futures we don't want, and tack away from them. We can also imagine the futures we do want, and awkwardly burl our logs toward them. As the future perfect turns into the present perfect, we can apply ourselves to creating a tolerable present and future—for ourselves and for the rest of life.

Current and future generations may yet find their way toward sanity. Sociologist Carrie Friese speculates that as the effects of climate change escalate, conservationists and others will be more and more motivated by a sense of multispecies solidarity—a profound understanding that, as Rachel Carson warned, we are "affected by the same environmental influences that control the lives of all the many thousands of other species." Indeed, the young climate activists who filled city streets around the world in 2019 are fighting for their own future; the Extinction Rebellion movement, launched in London in 2018, is rebelling against human extinction as well as the extinction of other species. The coronavirus pandemic most likely began when the international wildlife trade carried a virus from another species into human bodies. Our vulnerabilities are becoming more interconnected, and more immediate.

If we're lucky, this sense of solidarity will also overcome the impasses within the environmental and conservation movements—the generations-long arguments between utilitarians and preservationists, conservation-minded hunters and animal welfare advocates. Like the human societies they work within, these movements must constantly weigh individual interests against the common good, and those decisions are only becoming more difficult. Conservationists and environmentalists of all inclinations can influence one another for the better, pushing for solutions that minimize harm to individuals while maximizing benefits to populations and species, and they can and should find solidarity in their shared sympathy for the rest of life. Love of other species is fostered by individual connections between humans and their fellow animals; the determination to prevent extinction—and protect abundance—is the desire to reduce suffering, multiplied.

Our emotional bonds with other species and their members, so fundamental to conservation, still form in what Aldous Huxley called "the world of direct apprehension." Crouching over puddles to watch tadpoles wriggle toward metamorphosis remains the classic childhood lesson in the wonders and vulnerabilities of all species—our own included. And as amphibian lives, like human lives, are ever more disrupted by climate change and disease, we have all the more reason to take the morals of their stories to heart.

What a piece of work we are. Darwin, riffing on *Hamlet* and anticipating Aldous Huxley, concluded *The Descent of Man* with the observation that "man with all his noble qualities, with sympathy which feels for the most debased, with benevolence which extends not only to other men but to the humblest living creature . . . with all these exalted powers—Man still bears in his bodily frame the indelible stamp of his lowly origin."

We are not as gods but as frogs, and we had better get good at it.

Acknowledgments

Books, like beasts, depend on relationships, and I'm grateful to all who so generously helped me bring this book to life.

Thank you to the Alfred P. Sloan Foundation and its Public Understanding of Science and Technology program, whose support allowed me to dig deeper and travel farther than I otherwise would have been able to. I am also indebted to the Alicia Patterson Foundation, whose fellowship in 2011 gave me the chance to begin thinking about the human history of species conservation.

Much gratitude to my agent, Mollie Glick, who believed in me and this project and found us a welcoming home at W. W. Norton, and to my editor, Matt Weiland, who was a source of wise counsel, good cheer, and steady encouragement from first draft to finished manuscript.

Thank you to Zarina Patwa, Lily Gellman, and Allegra Huston, who improved these pages with their attention and insights; to Sarahmay Wilkinson, who designed a perfect cover; to Erin Sinesky Lovett and Steve Colca, who made sure that this book found its way from my hands to yours; to Margaret Burke, for her help with image research; and to fact-checker Emily Krieger, who expertly cleared away my nonsense.

Librarians and archivists make the world a better place, and I'm especially thankful for those at the Special Collections at the University of Arizona Libraries, the Library of Congress, the Hawk Mountain Sanctuary Archives, the Lilly Library at Indiana University at Bloomington, the New York Public Library Manuscripts and Archives Division, the Smithsonian Institution Archives, the University of Wisconsin–Madison Archives, and the Sam Cohen Library in Swakopmund, Namibia. Thanks also to the interlibrary loan staff of the Fort Vancouver Regional Libraries, who patiently tracked down any number of obscure books and documents on my behalf, and to Abby Dockter and Michelle Winglee, who helped me search archives from afar.

Many, many people helped me understand the past and present of conservation, and introduced me to their own beloved beasts. Particular thanks to George Archibald, Keith Aune, Karen Becker, Rich Beilfuss, Jim and Dottie Brett, Ervin Carlson, Larry Cook, Carl Cotter, Charlie and Betty Crow Chief, Robert DeCandido (better known in Central Park as "Birding Bob"), Lesley DeFalco, Betsy Didrickson, Deborah Edgemann, Paul Ehrlich, Todd Esque, Carrie Friese, Emmanuel Frimpong, Laurie Goodrich, Steve Gotte, Boas Hambo, Sonia Hambo, Lyle Heavy Runner, Margaret Jacobsohn, David Johns, Bob Kaplan, Keith Karoly, John Kasaona, Edison Kasupi, Cheri Kicking Woman, Anne Lacy, William Laurence, Estella Leopold Jr., Simon Mahood, Michael Mascia, Char Miller, Jeff Muntifering, Dave Parchen, Suzanne Peurach, Kent Redford, Oliver Ryder, Rafe Sagarin, Cecil Schwalbe, Kim Smith, Bill Snape, Kelly Stoner, Steve Swenson, Stanley Temple, Chastedy Williamson-Tatsey, Mark Vermeij, Marianne Wellington, Carol and Victor Yannacone, and Erika Zavaleta.

Thank you to Curt Meine, who graciously shared his reflections on conservation, and his deep knowledge of Aldo Leopold, over the course of this project; to Marcia Bjornerud, who treated me to a tour of her beloved bedrock; and to Ellen Lewis, who provided cross-genre inspiration and the epigraph.

Thank you to *The Atlantic, bioGraphic, High Country News,* and *The Last Word on Nothing,* which published earlier versions of some passages.

Thank you to my alma mater, Reed College, which supplied access to the JSTOR database (more important than you might think) and to its well-loved ski cabin on Mount Hood, known to me as Scrounger Yaddo.

I am obliged to those who took the time to read various drafts of these chapters, and better them with their smarts and expertise: Ian Billick, Jonathan Cobb, Mary Ellen Hannibal, Josh Howe, Ann Finkbeiner, Suzanne Londeree, Ben Minteer, and Richard Pyle. Thank you, all.

For writerly solidarity, and for crucial advice and assistance of many kinds, I'm grateful to Paolo Bacigalupi, Cynthia Barnett, Juli Berwald, Craig Childs, Alexandra Elbakyan, Eileen Garvin, Kate Greene, Lisa Hamilton, Greg Hanscom, Tom Hayden, Lisa Jones, Emma Marris, Sierra Crane Murdoch, Lauren Oakes, Andrea Pitzer, Gabe Ross, Ella Taylor, Kevin Taylor, JT Thomas, Craig Welch, Amy Williams, Florence Williams, and Ed Yong; to the members of Scilance, Babylance, and the Slackline; to my past and present colleagues at *The Atlantic* and *High Country News*; and to Judith Lewis Mernit, who five years ago sat me down on a bench by the River Thames and more or less ordered me to write this book. To my friends who spend their days doing something other than writing, thank you so much for listening; I owe you.

While I was working on this project, conservation lost two of its greats: Michael Soulé and Garth Owen-Smith, both of whom passed away in 2020. I wish they had met each other; their all-night argument by the campfire would have been something to hear. Each was a catalyst for change, and soon, I hope, their ideas will converge. I was lucky to learn from them both.

Finally, heartfelt thanks to my parents, Margaret and Rolf Nijhuis, and my in-laws, Joanne and Wes Perrin, for their confidence in me during this adventure. Thanks to Pika, Ziggy, and Stardust, my rascally domesticated beasts, for the comic relief. And thank you to Jackson Perrin and Sylvia Perrin, my best beloveds, for everything.

Notes

INTRODUCTION: AESOP'S SWALLOWS

1 **When the swallow first noticed the mistletoe:** Adapted from Aesop, "The House-martin and the Birds," Fable 349 in *The Complete Fables*, translated by Olivia and Robert Temple (New York: Penguin, 1998), 255.

1 **Almost nothing is known about Aesop himself:** While we can only speculate about Aesop's life, the most detailed speculation is found in the multiple versions of the anonymously authored "Aesop Romance." The version I consulted was translated by Lloyd Daly and included in *The Anthology of Ancient Greek Popular Literature*, edited by William Hansen (Bloomington: Indiana University Press, 1998), 106–62.

2 **they are animal enough:** Jeremy B. Lefkowitz, "Aesop and Animal Fable," in *The Oxford Handbook of Animals in Classical Thought and Life*, edited by Gordon Lindsay Campbell (Oxford: Oxford University Press, 2014), 1–23.

4 **The Bible said:** All biblical references are from the King James Bible.

6 **as geographer William Adams observes:** William M. Adams, *Against Extinction: The Story of Conservation* (London: Earthscan, 2004), xii–xiii.

6 **the last Yangtze river dolphin:** Samuel T. Turvey et al., "First Human-Caused Extinction of a Cetacean Species?," *Biology Letters* 3, no. 5 (2007): 537–40. The last documented sighting was in 2002.

6 **the last two northern white rhinos:** Sarah Gibbens, "After Last Male's Death, Is the Northern White Rhino Doomed?," *National Geographic*, March 20, 2018.

6 **the soon-to-be-last vaquita:** Ben Goldfarb, "The Endling: Watching a Species Vanish in Real Time," *Pacific Standard*, June 8, 2018.

7 **An estimated one million species are now threatened with extinction:** Sandra Diaz

et al., "Pervasive Human-Driven Decline of Life on Earth Points to the Need for Transformative Change," *Science* 366 (2019): eaax3100.

7   **at least eighteen hundred people were murdered**: Nathalie Butt et al., "The Supply Chain of Violence," *Nature Sustainability* 2 (2019): 742–47. Statistics for 2018 and 2019, like those analyzed by Butt et al., were compiled by Global Witness, globalwitness.org.

8   **The Indian emperor Ashoka**: Shravasti Dhammika, *The Edicts of King Asoka: An English Rendering* (Colombo, Sri Lanka: Buddhist Publication Society, 1993), 39. The emperor's name can be transliterated from Sanskrit as Ashoka or Asoka.

8   **Marco Polo reported that**: *The Travels of Marco Polo, Book 2*, translated by Henry Yule, edited by Henri Cordier (London: John Murray, 1920), chapter 20.

8   **"the oldest task in human history"**: Aldo Leopold, "Engineering and Conservation," in *Aldo Leopold: A Sand County Almanac and Other Writings on Ecology and Conservation*, edited by Curt Meine (New York: Library of America, 2013), 410.

9   **(and various orthographies)**: Early conservationists were advocates for "wild life" or "wild-life"; the World Wildlife Fund was originally called the World Wild Life Fund. For a discussion of the connotations of these variations, see Etienne Benson, "From Wild Lives to Wildlife and Back," *Environmental History* 16, no. 3 (2011): 418–22.

9   **"That the situation is hopeless"**: Aldo Leopold to William Vogt, undated response to a January 21, 1946, letter from Vogt, Aldo Leopold Papers, University of Wisconsin Archives, series 10–2, box 4, folder 11, 911. Reprinted in *Leopold: A Sand County Almanac and Other Writings*, 830–31.

CHAPTER ONE: THE BOTANIST WHO NAMED THE ANIMALS

Unless otherwise noted, quotations from Carl Linnaeus's memoirs in this chapter are from Wilfrid Blunt, *Linnaeus: The Compleat Naturalist* (London: Frances Lincoln, 2004).

10   **This taxonomic tradition**: Joeri Witteveen, "Suppressing Synonymy with a Homonym: The Emergence of the Nomenclatural Type Concept in Nineteenth Century Natural History," *Journal of the History of Biology* 49 (2016): 135–89.

11   **Now there are sixteen**: Kathlene L. Joyce et al., "Phylogeography of the Slimy Salamander Complex (*Plethodon*: Plethodontidae) in Alabama," *Copeia* 107, no. 4 (2019): 701–7.

11   **the most destructive pathogen on earth**: Ben C. Scheele et al., "Amphibian Fungal Panzootic Causes Catastrophic and Ongoing Loss of Biodiversity," *Science* 363 (2019): 1459–63. For a critique of this study's methods and a response by the authors, see Max R. Lambert et al., *Science* 367 (2020): eaay1838, and Ben C. Scheele et al., *Science* 367 (2020): eaay2905.

12   **the species called attention to itself by going missing**: Ron Santiago et al., "Popu-

lation Decline of the Jambato Toad *Atelopus ignescens* (Anura: Bufonidae) in the Andes of Ecuador," *Journal of Herpetology* 37, no. 1 (2003): 116–26.

12   **when a young boy discovered a tiny remnant population:** Lou Del Bello, "Boy Finds 'Extinct' Frog in Ecuador," *New Scientist*, July 7, 2017. The boy and his family were awarded $1000 by the Jambatu Center for Amphibian Research and Conservation, an Ecuadorian organization that had offered the prize to the first person to find an *A. ignescens* specimen.

12   **a population is generally defined:** For an overview of the historical debate and confusion surrounding this term, see L. H. M. Jonckers, "The Concept of Population in Biology," *Acta Biotheoretica* 22, no. 2 (1973): 78–108.

12   **"Frogs are strange creatures":** Henry David Thoreau, *The Journal 1837–1861* (New York: New York Review Books, 2009), 492.

13   **elaborate local taxonomies are still in service:** Some of these taxonomies are described in Carol Yoon, "A Surprise in the Tower of Babel," chapter 5 in *Naming Nature: The Clash Between Instinct and Science* (New York: W. W. Norton, 2009), 117–45.

13   **"find out what the animals themselves know":** Quoted in Richard Conniff, "When Animals Attack Our Attempts to Categorize Them," *Discover*, November 18, 2010.

13   **We will never know what the animals know:** For a famous meditation on this idea, see Thomas Nagel, "What Is It Like to Be a Bat?" *Philosophical Review* 83, no. 4 (1974): 435–50.

13   **the estimated nine million species on earth:** This figure is based on an estimate of eukaryotic species by Camilo Mora et al., "How Many Species Are There on Earth and in the Ocean?" *PLOS Biology* 9, no. 8 (2011): e1001127. When prokaryotes—the domains Bacteria and Archaea—are included in the total, estimates increase to as many as a trillion species. See Kenneth J. Locey and Jay T. Lennon, "Scaling Laws Predict Global Microbial Diversity," *Proceedings of the National Academy of Sciences* 113, no. 21 (2016): 5970–75. As for the number of formally named species, the 2019 checklist of the Catalogue of Life, a database of species databases, estimates that two million eukaryotic species are known to taxonomists.

14   **"From the depths of his heart":** Sten Lindroth, "The Two Faces of Linnaeus," in *Linnaeus: The Man and His Work*, edited by Tore Frängsmyr (Berkeley: University of California Press, 1983), 25.

14   **"first view of the conveniences of arrangement":** Florence Caddy, *Through the Fields with Linnaeus: A Chapter in Swedish History*, vol. 1 (London: Longmans, Green, 1887), 49.

14   **the haphazard naming of these plants and animals:** The consequences of this haphazard naming are recounted by Andrea Wulf in "The Bright Beam of Gardening," chapter 2 in *The Brother Gardeners: Botany, Empire, and the Birth of An Obsession* (New York: Viking, 2010), 34–47. Wulf quotes seventeenth-century apothecary and botanist John Parkinson, who warned readers of his plant catalog that "scarce one

of twentie of our Nursery Men doe sell the right, but give one for another" and "cheate men of their money."

17 **invisibly embedded into every scientific name**: Daniel Merriman, "Peter Artedi: Systematist and Ichthyologist," *Copeia* 1938, no. 1 (1938): 33–39.

18 **"two husbands in the same marriage"**: Quoted in William T. Stearn, "Appendix I: Linnaean Classification, Nomenclature, and Method," in Wilfrid Blunt, *Linnaeus: The Compleat Naturalist* (London: Frances Lincoln, 2004), 259.

18 **"nothing could equal the gross prurience"**: Stearn, "Appendix I," 261.

18 **the nested hierarchy Artedi had established for fish**: Daniel Merriman, "Peter Artedi: Systematist and Ichthyologist," *Copeia* 1938, no. 1 (1938): 33–39.

19 **"elementary order for an era of anarchy"**: Donald Worster, *Nature's Economy: A History of Ecological Ideas*, 2nd ed. (Cambridge: Cambridge University Press, 1994), 32.

19 **a taxonomic fallacy that haunts us still**: Jonathan Marks, "Long Shadow of Linnaeus's Human Taxonomy," letter to the editor, *Nature* 447 (2007): 28.

20 **known to science by the same names Linnaeus gave them**: My thanks to Richard Pyle of the Bernice Pauahi Bishop Museum and Markus Döring of the Global Biodiversity Information Facility for providing this figure.

20 **"Ordinary languages grow spontaneously"**: J. Chester Bradley, preface to the first edition of the International Code of Zoological Nomenclature, 1961. Quoted in the introduction to the current code, iczn.org/the-code.

20 **it didn't start keeping track**: Richard Pyle and Ellinor Michel, "ZooBank: Developing a Nomenclatural Tool for Unifying 250 Years of Biological Information," *ZooTaxa* 1950 (2008): 39–50. ZooBank became the official register of zoological nomenclature under the ICZN in 2012.

20 **Charles Davies Sherborn, a London bookseller**: Neal Evenhuis, "Charles Davies Sherborn and the 'Indexer's Club,'" *ZooKeys* 550 (2016): 13–32.

20 **Contemporary taxonomists can't hope for the glory**: Peter Whitehead, "Systematics: An Endangered Species," *Systematic Zoology* 39, no. 2 (1990): 179–84. Taxonomy, the science of naming species, is a branch of systematics, the study of evolutionary relationships among species.

21 **"the description and mapping of the world biota"**: Edward O. Wilson, "On the Future of Conservation Biology," *Conservation Biology* 14, no. 1 (2000): 1.

21 **"If you do not know the names of things"**: Carl Linnaeus, *Linnaeus' Philosophia Botanica*, translated by Stephen Freer (Oxford: Oxford University Press, 2003), 169.

21 **"For in order to care deeply about something important"**: Wilson, "Future of Conservation Biology," 2.

21 **A related toad, identified in 2004**: Stefan Lötters et al., "A New and Critically Endangered Species of *Atelopus* from the Andes of Northern Peru (Anura: Bufonidae)," *Revista Española de Herpetología* 18 (2004): 101–9.

21 **"most specially greedy, strong and wicked"**: Michael F. Bates, "A Taxonomic Revision of the South-eastern Dragon Lizards of the *Smaug warreni* (Boulenger) Species

Complex in Southern Africa, with the Description of a New Species (Squamata: Cordylidae)," *PeerJ* 8 (2020): e8526. Quote from J. R. R. Tolkien, *The Hobbit, or There and Back Again* (Boston: Houghton Mifflin, 1978), 28.

21 **three species of soft-nosed chameleons**: David Prötzel et al., "Untangling the Trees: Revision of the *Calumma nasutum* Complex (Squamata: Chameleonidae)," *Vertebrate Zoology* 70, no. 1 (2020): 23–59.

21 **the praying mantis *Vates phoenix***: Julio Rivera et al., "A New Species and First Record of *Vates* Burmeister, 1838 from the Atlantic Rainforest (Mantodea: Vatinae)," *European Journal of Taxonomy* 598 (2020): 1–25.

22 **they named it the Cambodian tailorbird**: Simon Mahood et al., "A New Species of Lowland Tailorbird (Passeriformes: Cisticolidae: *Orthotomus*) from the Mekong Floodplain of Cambodia," *Forktail* 29 (2013): 1–14.

22 **horsefly named after Beyoncé**: Brian Lessard, "New Species of the Australian Horse Fly Subgenus *Scaptia* (*Plinthina*) Walker 1850 (Diptera: Tabanidae), Including Species Descriptions and a Revised Key," *Australian Journal of Entomology* 50, no. 3 (2011): 241–52.

22 **a spider after David Bowie**: Peter Jäger, "Revision of the Huntsman Spider Genus *Heteropoda* Latreille 1804: Species with Exceptional Male Palpal Conformations from Southeast Asia and Australia (Arachnida, Araneae: Sparassidae: Heteropodinae)," *Senckenbergiana Biologica* 88 (2008): 239–310.

22 **a termite after . . . Fernando Botero**: Daniel Castro et al., "*Rustitermes boteroi*, a New Genus and Species of Soldierless Termites (Blattodea, Isoptera, Apicotermitinae) from South America," *ZooKeys* 922 (2020): 35–49.

22 **a tiny neon-purple fish**: Yi-Kai Tea et al., "*Cirrhilabrus wakanda*, a New Species of Fairy Wrasse from Mesophotic Ecosystems of Zanzibar, Tanzania, Africa (Teleostei, Labridae)," *ZooKeys* 863 (2019): 85–96.

22 **a millimeter-long tropical beetle**: Michael Darby, "Studies of Ptiliidae (Coleoptera) in the Spirit Collection of the Natural History Museum, London, 6: New Species and Records Collected by W. C. Block in Kenya and Uganda, 1964–1965," *Entomologist's Monthly Magazine* no. 4 (2019): 239–57.

22 **"likely to give offence"**: Appendix A: Code of Ethics, iczn.org/the-code.

22 **species of slime-mold beetles**: Kelly B. Miller and Quentin Wheeler, "Slime-Mold Beetles of the Genus *Agathidium* Panzer in North and Central America: Coleoptera, Leiodidae, Part 2," *Bulletin of the American Museum of Natural History*, no. 291 (2005): 143. Wheeler, whose team also named an *Agathidium* beetle after then Secretary of Defense Donald Rumsfeld, said at the time that the tributes were strictly in homage: "We admire these leaders as fellow citizens who have the courage of their convictions," he said. President Bush called to thank him.

22 **a moth with feathery yellow head scales**: Vazrick Nazari, "Review of *Neopalpa* Povolný, 1998 with Description of a New Species from California and Baja California, Mexico (Lepidoptera, Gelechiidae)," *ZooKeys* 646 (2017): 79–94.

22   **one eyeless, cave-dwelling Slovenian beetle species**: May Berenbaum, "Ice Break-
     ers," *American Entomologist* 56, no. 3 (2010): 132–33.

23   **Linnaeus, who had named the species**: William T. Stearn, "The Background of Lin-
     naeus's Contributions to the Nomenclature and Methods of Systematic Biology,"
     Systematic Zoology 8, no. 1 (1959): 4–22.

23   **To them, the metaphorical tree of life**: Mark A. Ragan, "Trees and Networks Before
     and After Darwin," *Biology Direct* 4 (2009): 43.

23   **"*Natura non facit saltus*"**: Quoted in Ernst Mayr, *The Growth of Biological Thought:
     Diversity, Evolution, and Inheritance* (Cambridge, MA: Belknap Press of Harvard
     University Press, 1982), 259.

23   **Linnaeus apparently became less sure about this**: Mayr, *The Growth of Biological
     Thought*, 259.

23   **"so connected, so chained together"**: "The Economy of Nature, a Dissertation Pre-
     sided over by Carl Linnaeus," translated by Benjamin Stillingfleet, in *The History
     and Philosophy of Science*, edited by Daniel McKaughan and Holly VandeWall (Lon-
     don: Bloomsbury, 2018), 759. At the time, academic theses were written by super-
     visors and defended by their students. See Geir Hestmark, "*Oeconomia Naturae* L."
     Nature 405 (2000): 19.

24   **represented evolutionary history as a gnarly oak**: See Benoît Dayrat, "The Roots of
     Phylogeny: How Did Haeckel Build His Trees?" *Systematic Biology* 52, no. 4 (2003):
     515–27. Dayrat argues that while Haeckel was an admirer of Darwin and champi-
     oned his work, his trees reflect earlier ideas about a single line of "progress" from
     so-called lower to higher organisms.

25   **a previously unrecognized category of life**: Abby Olena, "Discovering Archaea,
     1977," *Scientist*, March 2014. In addition to the plant and animal kingdoms,
     Eukarya includes the kingdoms Protista and Fungi.

25   **a process called horizontal gene transfer**: Shannon M. Soucy et al., "Horizontal Gene
     Transfer: Building the Web of Life," *Nature Reviews Genetics* 16 (2015): 472–82.

25   **dozens of finer-grained definitions**: For an overview of the madness, see Susan
     Milius, "The Fuzzy Art of Defining Species: A Vital Concept Sparks Many Argu-
     ments," *Science News* 198, no. 2 (2017): 22–24.

25   **in a room of *n* biologists**: John Wilkins, "How Many Species Concepts Are There?"
     Guardian, October 20, 2010.

25   **there is no ultimate authority on what constitutes a species**: Stephen T. Garnett and
     Les Christidis, "Taxonomy Anarchy Hampers Conservation," Comment, *Nature*
     546 (2017): 25–27. This critique was emphatically countered by Scott A. Thomson
     et al., "Taxonomy Based on Science Is Necessary for Global Conservation," *PLoS
     Biology* 16, no. 3 (2018): e2005075.

26   **there are four species of giraffe, not just one**: Julian Fennessey et al., "Multi-Locus
     Analyses Reveal Four Giraffe Species Instead of One," *Current Biology* 26, no. 18
     (2016): 2543–49.

26 **reduce the number of recognized tiger subspecies**: Andreas Wilting, "Planning
    Tiger Recovery: Understanding Intraspecific Variation for Effective Conservation,"
    Science Advances 1, no. 5 (2015): e1400175.

26 **Some are more endangered than their antecedents**: Ashley T. Simkins et al., "The
    Implications for Conservation of a Major Taxonomic Revision of the World's Birds,"
    Animal Conservation 23, no. 4 (2020): 345–52.

26 **species are the fundamental units**: E. O. Wilson, "The Fundamental Unit," chapter
    4 in *The Diversity of Life* (Cambridge, MA: Harvard University Press, 2010). While
    acknowledging the many complexities of the species concept, Wilson wrote: "Not to
    have a natural unit such as the species would be to abandon a large part of biology
    into free fall, all the way from the ecosystem down to the organism."

CHAPTER TWO: THE TAXIDERMIST AND THE BISON

Unless otherwise noted, Hornaday's recollections in this chapter are from his unpublished memoir *Eighty Fascinating Years*, William T. Hornaday Papers, Library of Congress, box 112, 5–6.

28 **other types of "nature study" were fashionable**: Sally Gregory Kohlstedt, "Nature,
    Not Books: Scientists and the Origins of the Nature-Study Movement in the 1890s,"
    Isis 96, no. 3 (2005): 324–52. Also see Mark V. Barrow Jr., "The Specimen Dealer:
    Entrepreneurial Natural History in America's Gilded Age," *Journal of the History of
    Biology* 33 (2000): 493–534.

29 **"about the biggest thing ever attempted"**: Harry Godwin, *Washington Star*, March
    10, 1888, reprinted in William Hornaday, "The Extermination of the American
    Bison," in *Report of the National Museum 1886–1887*, 547.

29 **the tribe's reservation in northwestern Montana**: The Blackfeet reservation in Mon-
    tana, established by treaty with the U.S. government in 1855, is home to the Black-
    feet Nation. In other words, the reservation is the land; the Blackfeet Nation is the
    political entity.

29 **Both subspecies of *Bison bison***: The taxonomy of *Bison bison* has a tangled history. See
    C. Cormack Gates et al., *American Bison: Status Survey and Conservation Guidelines*
    (Gland, Switzerland: IUCN, 2010). Today, *B. b. bison* and *B. b. athabascae* are gener-
    ally recognized as subspecies, but for a dissenting opinion, see Matthew Cronin et al.,
    "Genetic Variation and Differentiation of Bison (*Bison bison*) Subspecies and Cattle
    (*Bos taurus*) Breeds and Subspecies," *Journal of Heredity* 104, no. 4 (2013): 500–09.

30 **one of four Blackfoot-speaking tribes**: Blackfoot is the language spoken by the four
    tribes and First Nations of the Blackfoot Confederacy, one of which is the Blackfeet,
    or Southern Piegan. The other three are Piikani, or Northern Piegan; the Kainai,
    or Blood; and the Siksika, or Blackfoot. The traditional territory of the confederacy
    extends from northern Montana into southern Saskatchewan and Alberta.

31 **bison numbers rose and fell**: Geoff Cunfer, "The Decline and Fall of the Bison Empire," in *Bison and People on the North American Great Plains: A Deep Environmental History*, edited by Geoff Cunfer and Bill Waiser (College Station: Texas A&M University Press, 2016), 1.

31 **European settlers encountered the bison population at its peak**: Cunfer, "Decline and Fall," 10.

31 **an estimated twenty to thirty million bison**: Andrew C. Isenberg, *The Destruction of the Bison: An Environmental History, 1750–1920* (Cambridge: Cambridge University Press, 2000), 25. See Cunfer, "Decline and Fall," for further discussion.

31 **George Washington and a few companions**: George Washington, "2 November 1770," *The Diaries of George Washington*, vol. 2, *1766–70* (Charlottesville: University Press of Virginia, 1976), 307–8. "We proceeded up the River with the Canoe about 4 Miles more, & then incampd & went a Hunting: killd 5 Buffaloes & wounded some others—three deer&ca. This country abounds in Buffalo & wild game of all kinds." Available online at the Library of Congress.

31 **a "moving multitude" of bison**: Quoted in Mark V. Barrow Jr., *Nature's Ghosts: Confronting Extinction from the Age of Jefferson to the Age of Ecology* (Chicago: University of Chicago Press, 2009), 93.

32 **"confine the Indians to smaller areas"**: Quoted in Isenberg, *Destruction of the Bison*, 152. The much-repeated story that General Philip Sheridan, in 1875, applauded commercial bison hunters for "destroying the Indian's commissary" is convincingly debunked by Dan Flores, "Reviewing an Iconic Story: Environmental History and the Demise of the Bison," in *Bison and People*, 36.

32 **fewer than three hundred free-roaming plains bison**: Hornaday, "Extermination of the American Bison," 529.

32 **once among the richest societies in the world**: Donna Feir et al., "The Slaughter of the Bison and Reversal of Fortunes on the Great Plains," University of Victoria Economics Department Discussion Paper, July 9, 2017.

34 **already extinguished dozens of bird species**: Stuart Pimm et al., "Human Impacts on the Rates of Recent, Present, and Future Bird Extinctions," *Proceedings of the National Academy of Sciences* 103, no. 29 (2006): 10941–46. Pimm et al. conclude that the forty-three bird extinctions recorded between 1500 and 1800 amount to "a considerable underestimate."

34 **studies of elephant skulls**: The story of Cuvier and his living and extinct proboscideans is told by Elizabeth Kolbert in "The Mastodon's Molars," chapter 2 in *The Sixth Extinction: An Unnatural History* (New York: Henry Holt, 2014).

34 **"Such is the economy of nature"**: Thomas Jefferson, *Notes on the State of Virginia* (Boston: Lily and Wait, 1832), 52. Available online at the Library of Congress. The animals Jefferson refers to as "mammoths" have been called mastodons since 1806, when Cuvier finally named the species he had distinguished from the mammoth ten years earlier.

34   **"rare or extinct" animals**: Jefferson, "Instructions for Meriwether Lewis," June 20, 1803, in *The Papers of Thomas Jefferson, Vol. 40*, edited by James P. McClure (Princeton: Princeton University Press, 2013), 178.

34   **"We need not marvel at extinction"**: Charles Darwin, *On the Origin of Species*, reprint of 1859 edition (New York: Signet Classics, 2003), 353. Though Darwin published five revisions of *Origin*, the last one in 1872, he never altered this line. He did become somewhat less sanguine about extinction in later years, especially when he contemplated the possibility of human extinction. See Gillian Beer, "Darwin and the Uses of Extinction," *Victorian Studies* 51, no. 2 (2009): 321–31.

35   **"the nobler animals"**: Henry David Thoreau, *The Journal 1837–1861* (New York: New York Review Books, 2009), 371.

35   **"The myriad forms of animal and vegetable life"**: George Perkins Marsh, *Man and Nature*, edited by David Lowenthal (Seattle: University of Washington Press, 2003), 18.

35   **one of the first naturalists to recognize**: German naturalist Georg Heinrich von Langsdorff provided an early hint in 1805, when he reported that the "mysterious and curious" Steller's sea cow, a giant relative of the manatee once abundant in the North Pacific, had been hunted out of "present creation." See Ryan Tucker Jones, "A 'Havock Made among Them': Animals, Empire, and Extinction in the Russian North Pacific, 1741–1810," *Environmental History* 16, no. 4 (2011): 585–609.

35   **"the exterminating process"**: Quoted in Henry M. Cowles, "A Victorian Extinction: Alfred Newton and the Evolution of Animal Protection," *British Journal for the History of Science* 46, no. 4 (2013): 701.

35   **"I believe it may be affirmed with confidence"**: T. H. Huxley, address to the Inaugural Meeting of the Fishery Congress (London: W. Clowes and Sons, 1883), 14.

35   **"The bison has had his day"**: "The American Bison," *San Francisco Chronicle*, February 9, 1890, 4.

36   **"sickening scenes of slaughter"**: Quoted in Isenberg, *Destruction of the Bison*, 145.

36   **"Having found out what you have to do"**: William Mathews, *Getting on in the World; Or, Hints on Success in Life* (Chicago: S. C. Griggs, 1879), 350.

37   **"the great game of the Empire"**: "The Dying Fauna of an Empire," *Saturday Review of Politics, Literature, Science and Art*, November 24, 1906, 635–36. According to the article, Salisbury's sudden awakening took place around 1896.

37   **"All inquiries elicited the same reply"**: Hornaday, "Extermination of the American Bison," 531.

38   **"It is a fearful job"**: Expedition journal entry, October 20, 1886, William T. Hornaday Papers, Library of Congress, box 112.

38   **"A little bit of Montana"**: Godwin, *Washington Star*, reprinted in Hornaday, "Extermination of the American Bison," 546.

38   **"the charms of wild nature"**: Quoted in James Andrew Dolph, "Bringing Wildlife to Millions: William Temple Hornaday, The Early Years 1854–1896," PhD diss., University of Massachusetts, 1975, 490.

39   "It now seems necessary for us to assume": William T. Hornaday to G. Brown Goode, December 2, 1887, quoted in Dolph, "Bringing Wildlife to Millions," 575. For more on the conservation intentions of early zoos, see Jeffrey Stott, "The Historical Origins of the Zoological Park in American Thought," *Environmental Review* 5, no. 2 (1981): 52–65.

40   "The day is fine": William T. Hornaday to Josephine Hornaday, June 19, 1896, William T. Hornaday Papers, Library of Congress, box 1.

40   an "alien city": Quoted in Jonathan Peter Spiro, *Defending the Master Race: Conservation, Eugenics, and the Legacy of Madison Grant* (Lebanon, NH: University of Vermont Press, 2009), 97.

40   "a world of clerks and teachers": William James, "The Moral Equivalent of War," in *Essays in Religion and Mortality* (Cambridge, MA: Harvard University Press, 1982), 162–73.

41   which he called the "Nordic race": Quoted in Miles A. Powell, *Vanishing America: Species Extinction, Racial Peril, and the Origins of Conservation* (Cambridge, MA: Harvard University Press, 2016), 97.

41   Adolf Hitler praised it as "my bible": Quoted in Spiro, *Defending the Master Race*, 357.

41   "conservation of that race": Madison Grant, *The Passing of the Great Race; Or, The Racial Basis of European History* (New York: Charles Scribner's Sons, 1916), ix. Grant revised his book three times, and the fourth edition, published in 1921, was nearly twice the size of the first. His editor was the renowned Maxwell Perkins, who worked with Ernest Hemingway and F. Scott Fitzgerald. See Spiro, *Defending the Master Race*, 161–66.

42   "this incident will form its most amusing passage": Quoted in Mitch Keller, "The Scandal at the Zoo," *New York Times*, August 6, 2006.

43   "intensely American stock": Quoted in Richard Slotkin, "Nostalgia and Progress: Theodore Roosevelt's Myth of the Frontier," *American Quarterly* 33, no. 5 (1981): 626.

43   "conflicts with the weaker race": Slotkin, "Nostalgia and Progress," 624.

44   "There will always be pigeons in books": Aldo Leopold, "On a Monument to the Pigeon," in *Aldo Leopold: A Sand County Almanac and Other Writings on Ecology and Conservation*, edited by Curt Meine (New York: Library of America, 2013), 97.

44   "It is now generally recognized as ethically wrong": Joseph Grinnell, "Bird Life as a Community Asset," *California Fish and Game* 1, no.1 (1914): 20.

45   "Director Hornaday of the Bronx Zoological Park": "Bison Preserves," *New York Times*, November 3, 1907, 8.

46   Only in 1866 had zoologist Ernst Haeckel: Donald Worster, *Nature's Economy: A History of Ecological Ideas*, 2nd ed. (Cambridge: Cambridge University Press, 1994), 192.

46   "the study of all those complex interrelations": Quoted in Carolyn Merchant, *The Columbia Guide to American Environmental History* (New York: Columbia University Press, 2002), 160.

46   **"entangled bank"**: Darwin, *On the Origin of Species*, 1859 edition reprint, 507. This
     famous phrase, from the poetic final passage of *On the Origin of Species*, is believed
     to have been inspired by a real place—a hillside near Darwin's home in southeast-
     ern England called Downe Bank, known for its profusion of orchids.

46   **In Yellowstone, bison still have more influence**: Chris Geremia et al., "Migrating
     Bison Engineer the Green Wave," *Proceedings of the National Academy of Sciences*
     116, no. 51 (2019): 25707–13.

49   **students serve out detention**: I am indebted to Amy Martin and her team at the
     Threshold podcast for this detail, which appears in "Oh, Give Me a Home," episode
     7 of season 1.

49   **"an important something"**: Lee Jones, a biologist with the U.S. Fish and Wildlife
     Service, said this during a talk at the American Bison Society meeting in Banff,
     Alberta, in 2016.

50   **a diffuse, largely leaderless movement**: For more on the bison restoration move-
     ment's origins and mission, see Eric W. Sanderson et al., "The Ecological Future
     of the North American Bison: Conceiving Long-Term, Large-Scale Conservation of
     Wildlife," *Conservation Biology* 22, no. 2 (2008): 252–66.

52   **"To my illustrious successor"**: "W. T. Hornaday Note about Bison Specimens,"
     MAH–44711, Smithsonian Institution Archives, siarchives.si.edu.

54   **an "elitist conspiracy"**: Stephen Fox, *The American Conservation Movement: John
     Muir and His Legacy* (Madison: University of Wisconsin Press, 1986), 110.

54   **the first international wildlife conservation treaty**: Kurk Dorsey, "Putting a Ceiling
     on Sealing: Conservation and Cooperation in the International Arena, 1909–1911,"
     Environmental History Review 15, no. 3 (1991): 27–45.

CHAPTER THREE: THE HELLCAT AND THE HAWKS

Rosalie Edge's recollections in this chapter, unless otherwise noted, are from her unpublished memoir *An Implacable Widow*, Hawk Mountain Sanctuary Archives.

55   **On the morning of October 29, 1929**: The details of Black Tuesday are from Freder-
     ick Lewis Allen, *Since Yesterday: The 1930s in America, September 3, 1929–September
     3, 1929* (New York: Bantam, 1965) and Gordon Thomas and Max Morgan-Witts, *The
     Day the Bubble Burst: A Social History of the Wall Street Crash of 1929* (New York:
     Penguin, 1980).

57   **as a *New Yorker* writer once put it**: Robert Lewis Taylor, "Oh, Hawk of Mercy!" *The
     New Yorker*, April 17, 1948, 31.

57   **According to the meeting minutes**: National Association of Audubon Societies,
     October 29, 1929, Annual Meeting Minutes, National Audubon Society Records,
     Manuscripts and Archives Division, New York Public Library, box A175.

58   **"Toward wild life the Italian laborer"**: Quoted in Adam Rome, "Nature Wars, Cul-

ture Wars: Immigration and Environmental Reform in the Progressive Era," *Environmental History* 13 (July 2008): 437.

59 **"I did not like your letter"**: Rosalie Edge to Maurice Broun, January 14, 1950, Hawk Mountain Sanctuary Archives.

59 **"a citizen-scientist and militant political agitator"**: Dyana Z. Furmansky, *Rosalie Edge, Hawk of Mercy: The Activist Who Saved Nature from the Conservationists* (Athens, GA: University of Georgia Press, 2009), 3.

60 **the mysteries of flight and migration**: Edge's questions are echoed by Gregg Mitman in "The Hope and Promise of Birds," *Environmental History* 12, no. 2 (2007): 343–45.

60 **Fossil evidence suggests**: David W. Steadman, "Prehistoric Extinctions of Pacific Island Birds: Biodiversity Meets Zooarchaeology," *Science* 267 (1995): 1123–31.

60 **"Formerly, a great many flamingoes"**: Quoted in Richard Grove, *Green Imperialism: Colonial Expansion, Tropical Island Edens and the Origins of Environmentalism, 1600–1860* (Cambridge: Cambridge University Press, 1996), 244.

60 **Since the 1500s, humans have entirely eliminated**: Based on the species in the "Extinct" category of the IUCN Red List of Threatened Species, iucnredlist.org.

61 **altered species distributions throughout Latin America**: Paul D. Haemig, "Aztec Emperor Auitzotl and the Great-Tailed Grackle," *Biotropica* 10, no. 1 (1978): 11–17. Also see Paul D. Haemig, "Introduction of the Great-Tailed Grackle (*Quiscalus mexicanus*) by Aztec Emperor Auitzotl: Provenance of the Historical Account," *The Auk* 129, no. 1 (2012): 70–75.

61 **"The head-dress of our women"**: Quoted in Émile Langlade, *Rose Bertin: The Creator of Fashion at the Court of Marie-Antoinette* (New York: Charles Scribner's Sons, 1913), 53.

61 **"revolution in architecture"**: Quoted in chapter 5 of Henriette Campan, *Memoirs of the Court of Marie Antoinette, Queen of France*, vol. 1 (Boston: L. C. Page, 1900).

61 **of the seven hundred hats whose trimmings he observed**: Frank M. Chapman, "Birds and Bonnets," letter to the editor, *Forest and Stream* 26, no. 5 (1886): 84.

62 **In 1912, when Hornaday sent a young ornithologist**: William Hornaday, *Our Vanishing Wildlife: Its Extermination and Preservation* (New York: Charles Scribner's Sons, 1913), 117–22.

62 **The great auk was last documented**: Jeremy Gaskell, *Who Killed the Great Auk?* (Oxford: Oxford University Press, 2000), 129–30.

62 **"For all practical purposes"**: Henry M. Cowles, "A Victorian Extinction: Alfred Newton and the Evolution of Animal Protection," *British Journal for the History of Science* 46, no. 4 (2013): 700.

62 **"and their thoughtless, stupid devotion"**: William Hornaday, "Woman the Juggernaut of the Bird World," *New York Times*, February 23, 1913, 76. Lead article in the "Women's Pages."

62   **Chapman gave an address called**: Thor Hanson, *Feathers: Evolution of a Natural Miracle* (New York: Basic Books, 2011), 189.

62   **Virginia Woolf, in response to such rebukes**: Letter to *The Woman's Leader*, July 23, 1890, reprinted in *The Diary of Virginia Woolf*, vol. 2, *1920–1924* (New York: Harcourt, 1978), 337.

64   **"I have met *Cornurus* and he is mine"**: The story of Chapman vs. the Carolina parakeet is told in Mark V. Barrow Jr., *Nature's Ghosts: Confronting Extinction from the Age of Jefferson to the Age of Ecology* (Chicago: University of Chicago Press, 2009), 105–6. For an example of present-day collecting ethics, see Christopher E. Filardi, "Why I Collected a Moustached Kingfisher," *Audubon*, October 7, 2015.

64   **"I do not protect birds"**: Quoted in Barrow, *Nature's Ghosts*, 141.

65   **Florence Merriam Bailey**: For a belated but welcome remembrance of Bailey, see Jonathan Wolfe, "Overlooked No More: Florence Merriam Bailey, Who Defined Modern Bird-Watching," *New York Times*, July 17, 2019.

65   **"refrain from wearing the feathers"**: Royal Society for the Protection of Birds, "Our History," rspb.org.uk.

65   **"Not a *very* severe self-denying ordinance"**: "Birds and Bonnets," *Punch*, October 26, 1889, 197.

65   **a man of contradictions**: Gregory Nobles, "The Myth of John James Audubon," *Audubon*, July 31, 2020.

66   **"bordering on phrenzy"**: Quoted in Richard Rhodes, *John James Audubon: The Making of an American* (New York: Alfred A. Knopf, 2004), 22.

66   **"My best friends"**: John James Audubon, *The Life of John James Audubon, the Naturalist*, edited by Lucy Audubon (New York: G. P. Putnam's Sons, 1875), 93–94.

66   **"A magic power transported us"**: Quoted in Rhodes, *John James Audubon*, 279.

66   **"discovery of a New World"**: William Hornaday, chapter 2 in *Eighty Fascinating Years*, unpublished memoir, William T. Hornaday Papers, Library of Congress, box 112, 5–6.

67   **"an Association for the protection of wild birds"**: Quoted in Frank Graham Jr., *The Audubon Ark: A History of the National Audubon Society* (Austin: University of Texas Press, 1992), 9–13. For more on the role of women in the Audubon movement, see Carolyn Merchant, "George Bird Grinnell's Audubon Society: Bridging the Gender Divide in Conservation," *Environmental History* 15, no. 1 (2010): 3–30.

67   **"Fashion decrees feathers"**: Quoted in Graham, *Audubon Ark*, 13.

68   **"made themselves conspicuous"**: "'Dearie,' Say Women and Votes Kill an Anti-Plumage Measure," *Inter Ocean* (Chicago), April 8, 1910.

68   **In Florida in the early and mid-1920s**: Great egret counts submitted by Florida volunteers in 1921, 1922, 1925, 1926, 1927, from Christmas Bird Count database, audubon.org.

68 **During the Christmas Bird Count of 1937**: Report from Collier City, Florida, in "Bird-Lore's Thirty-Eighth Christmas Census," *Bird-Lore* 40 (1938): 45.

69 **"the first awakening of my mind"**: Quoted in Taylor, "Oh! Hawk of Mercy!," 34.

70 **"my earthquake"**: Rosalie Edge, "Winged Friendships," unpublished manuscript, Hawk Mountain Sanctuary Archives, 5.

71 **"Let us face facts now"**: Waldron DeWitt Miller et al., "A Crisis in Conservation," Hawk Mountain Sanctuary Archives, 1.

72 **"quixotic, truculent curiosity"**: Stephen Fox, *The American Conservation Movement: John Muir and His Legacy* (Madison: University of Wisconsin Press, 1986), 174.

72 **"worthy of a place with the Lamentations"**: Gilbert Pearson to Henry Fairfield Osborn, undated, National Audubon Society Records, Manuscripts and Archives Division, New York Public Library, box 179.

72 **"ascidians and isopods"**: Quoted in Furmansky, *Rosalie Edge*, 107.

73 **"No more pernicious nonsense"**: Willard Van Name, "Anti-Conservation Propaganda," letter to the editor, *Science* 61 (1925): 415.

73 **"the only honest, unselfish, indomitable hellcat"**: Taylor, "Oh! Hawk of Mercy!," 45.

73 **"fair and right to go after them with guns"**: W. T. Hornaday, reply to "Thin Out the Crows," a reader's letter in *Recreation* 21, no. 2 (1904): 113–14.

75 a **"sorehead"**: Mabel Osgood Wright to Willard Van Name, undated, National Audubon Society Records, Manuscripts and Archives Division, New York Public Library, box 179.

75 **"[We] could strike hard on any issue"**: Irving Brant, *Adventures in Conservation with Franklin D. Roosevelt* (Flagstaff, AZ: Northland Publishing, 1988), 17.

75 a **"common scold"**: Quoted in Furmansky, *Rosalie Edge*, 130.

75 **"The National Audubon Society recovered its virginity"**: Brant, *Adventures in Conservation with FDR*, 22.

76 **"Won't you ask Mrs. Edge"**: Quoted in Furmansky, *Rosalie Edge*, 234. This story is also recounted in Brant, *Adventures in Conservation with FDR*, 17, though Brant circumspectly attributes the quote to "a member of the President's cabinet."

76 **"Practically everything in the world is rentable"**: Quoted in Furmansky, *Rosalie Edge*, 168.

76 **"It is a job that needs some courage"**: Quoted in Furmansky, *Rosalie Edge*, 169.

76 **Guy Bradley, who guarded rookeries**: Graham, *Audubon Ark*, 50–58. The murder of Guy Bradley was the basis of "Plumes," a short story published in the *Saturday Evening Post* in 1930. The author was suffragist and Florida conservationist Marjory Stoneman Douglas, who would later write the influential book *The Everglades: River of Grass* (New York: Rinehart, 1947).

78 **"The time to protect a species"**: Quoted in Rosalie Edge, *An Implacable Widow*, unpublished memoir, Hawk Mountain Sanctuary Archives, 51. Edge used a version of this maxim on the Emergency Conservation Committee letterhead: "The time to

protect a species is while it is still common. The way to prevent the extinction of a species is never to let it become rare."

80 **"I was impressed with the great usefulness"**: Gilbert Pearson to Rosalie Edge, October 21, 1940, Hawk Mountain Sanctuary Archives.

81 **welcomed support from "sentimentalists"**: Quoted in Cowles, "A Victorian Extinction," 695–714.

81 **"Now the people have seen it"**: Quoted in Edge, *Implacable Widow*, 216.

82 **no exact transcript . . . survives**: David A. Greenwood, "Making Restoration History: Reconsidering Aldo Leopold's Arboretum Dedication Speeches," *Restoration Ecology* 25, no. 5 (2017): 681–88.

82 **"ecological destruction on a scale almost geological"**: Aldo Leopold, "The Arboretum and the University," in *Aldo Leopold: A Sand County Almanac and Other Writings on Ecology and Conservation*, edited by Curt Meine (New York: Library of America, 2013), 352–54.

CHAPTER FOUR: THE FORESTER AND THE GREEN FIRE

The voluminous Aldo Leopold Papers, housed in the University of Wisconsin Archives, are almost entirely digitized and freely available online.

84 **"Look, Mother, someone *lives* there!"**: Quoted in Estella B. Leopold, *Stories from the Leopold Shack: Sand County Revisited* (Oxford: Oxford University Press, 2016), 21.

84 **"All right, so now you *grow!*"**: Quoted in Estella B. Leopold, *Stories*, 29.

85 **"To keep every cog and wheel"**: Aldo Leopold, *Round River: From the Journals of Aldo Leopold* (Oxford: Oxford University Press, 1993), 147.

86 **"One of the penalties of an ecological education"**: Leopold, *Round River*, 165.

86 **"I am glad I shall never be young without wild country"**: Aldo Leopold, "Chihuahua and Sonora," in *Aldo Leopold: A Sand County Almanac and Other Writings on Ecology and Conservation*, edited by Curt Meine (New York: Library of America, 2013), 130.

86 **"His conviction was that conservation had to rest on a base"**: Curt Meine, *Correction Lines: Essays on Land, Leopold, and Conservation* (Washington, DC: Island Press, 2013), 125.

87 **The so-called "frontier line" was gone**: The census superintendent at the time was Robert Porter, a British journalist with no expertise in demography, and the 1890 census was riddled with problems (which continued even after the completion of the census; in 1921, most of the 1890 records were destroyed in a fire). Porter's influential pronouncement, often assumed to reflect a genuine transition in settlement patterns, was based on his general impression of rather spotty data. See Gerald D. Nash, "The Census of 1890 and the Closing of the Frontier," *Pacific Northwest Quarterly*, July 1980, 98–100.

87   "the meeting point between savagery and civilization": Frederick Jackson Turner, *The Frontier in American History* (New York: Henry Holt, 1921), 3. Turner first presented the "frontier thesis" in an 1893 lecture. The Leopold and Turner families were neighbors in Madison, and Turner's granddaughter helped plant pines at the shack. See Estella B. Leopold, *Stories*, 28.

87   "From my own observations I have found": Aldo Leopold, "The Wren," 1901 Theme Book (Burlington), Aldo Leopold Papers, University of Wisconsin Archives, series 10–13, box 3, folder 8. One of the few series in the archives that has not been digitized.

87   Leopold's first conservation mentor was his father: Aldo Leopold dedicated his book *Game Management* to his father, calling him a "Pioneer in Sportsmanship." The elder Leopold's hunting practices are recalled by his youngest son, Frederic, in "Recollections of an Old Member" (1977) and "Leopold Family Anecdotes" (1982), Aldo Leopold Papers, series 10–8, box 12, folder 4, 92 and 119–20 respectively.

88   He described sightings of more than a dozen bird species: Aldo Leopold to Clara Leopold, January 9, 1904, Aldo Leopold Papers, series 10–8, box 4, folder 2, 38–42.

88   "more than pleased with the country": Aldo Leopold to Clara Leopold, January 9, 1904, 41.

88   "I am now acquainted with 274 species": Aldo Leopold to Clara Leopold, May 18, 1904, Aldo Leopold Papers, series 10–8, box 4, folder 3, 417.

88   "baptizing species and describing feathers": Aldo Leopold, "Report of the Iowa Game Survey: Chapter One," Aldo Leopold Papers, series 10–3, box 5, folder 5, 735.

89   "of much interest and surprise": Aldo Leopold to Clara Leopold, February 1, 1906, Aldo Leopold Papers, series 10–8, box 5, folder 2, 365. Written during his first year at Yale.

89   "woods fever": Aldo Leopold to Carl Leopold, April 13, 1907, Aldo Leopold Papers, series 10–8, box 6, folder 1, 39.

89   "I do not want anyone with me but you": Theodore Roosevelt to John Muir, March 14, 1903, Theodore Roosevelt Digital Library, Dickinson State University, theodorerooseveltcenter.org.

90   "Mr. Roosevelt, when are you going to get beyond": Muir's (possibly self-serving) account of this conversation was recalled by Robert Underwood Johnson in his book *Remembered Yesterdays* (Boston: Little, Brown, 1923), 388.

90   "I used to envy the father of our race": John Muir, "Explorations in the Great Tuolomne Cañon," *Overland*, August 1873, 139–47.

90   "the greatest number, for the longest run": Gifford Pinchot, *Breaking New Ground* (Washington, DC: Island Press, 1998), 261–62. This description of the Forest Service's mission is from a letter that Pinchot ghostwrote for Secretary of Agriculture James Wilson.

90   "living society of living beings": Quoted in Char Miller, "The Greening of Gifford Pinchot," *Environmental History Review* 16, no. 3 (1992): 1–20.

90   **"seemed to have no right place in the landscape"**: John Muir, "The Passes," chapter 5 in *The Mountains of California*, vault.sierraclub.org.

91   **and struggled with his aversion to their unwashed faces**: Muir, "The Passes." For a qualified defense of Muir on this point, see Donald Worster, "John Muir and the Modern Passion for Nature," *Environmental History* 10, no. 1 (2005): 13. Also see Darryl Fears and Steven Mufson, "Liberal, Progressive—and Racist? The Sierra Club faces its White-Supremacist History," *Washington Post*, July 22, 2020.

91   **"for the benefit of all the people"**: Pinchot, *Breaking New Ground*, 476.

92   **"effort to reach an ever more complex understanding"**: Miller, "The Greening of Gifford Pinchot," 16.

92   **"will be as picturesque as though left wholly alone"**: Quoted in Char Miller, *Gifford Pinchot and the Making of Modern Environmentalism* (Washington, DC: Island Press, 2001), 95.

93   **"dreamers and theorists"**: Quoted in Pinchot, *Breaking New Ground*, 299–300.

93   **"turned handsprings in their wrath"**: Quoted in Pinchot, *Breaking New Ground*, 299–300.

93   **"American foresters trained by Americans in American ways"**: Quoted in Susan Flader, *Thinking Like a Mountain: Aldo Leopold and the Evolution of an Ecological Attitude Toward Deer, Wolves, and Forests* (Madison: University of Wisconsin Press, 1994), 9.

93   **"I am getting narrow as a clam"**: Quoted in Curt Meine, *Aldo Leopold: His Life and Work* (Madison: University of Wisconsin Press, 2010), 79.

94   **"I'd rather be a Supervisor"**: Quoted in Meine, *Aldo Leopold*, 80.

94   **"chaotic mass of very precipitous and rocky hills"**: Quoted in Meine, *Aldo Leopold*, 88. This observation was made by W. H. B. "Whiskey High Ball" Kent, one of the "boundary men" who surveyed proposed reserves for the U.S. Bureau of Forestry and, later, the Forest Service.

94   **"In the early morning a silvery veil hangs"**: Aldo Leopold to Clara Leopold, May 9, 1910, Aldo Leopold Papers, series 10–8, box 7, folder 2, 104.

95   **"I have consistently *ignored* the absence of letters"**: Aldo Leopold to Marie Leopold, August 8, 1911, Aldo Leopold Papers, series 10–13, box 2, folder 3.

96   **"It is the way of Americans to feel"**: Hornaday, *Our Vanishing Wildlife: Its Extermination and Preservation* (New York: Charles Scribner's Sons, 1913), 5.

96   **"North America, in its natural state"**: Aldo Leopold, "Game and Fish Handbook for Officers of the National Forests of Arizona and New Mexico," Aldo Leopold Papers, series 10–11, box 1, folder 4, 434.

97   **"To Mr. Aldo Leopold"**: Quoted in Meine, *Aldo Leopold*, 549, note 11.

97   **"to restore to every citizen his inalienable right"**: Quoted in Meine, *Aldo Leopold*, 161.

97   **"I do not know whether I have twenty days"**: Aldo Leopold to A. C. Ringland, February 14, 1916, quoted in Flader, *Thinking Like a Mountain*, 13.

97   "my dear Colonel": William Hornaday to Theodore Roosevelt, January 13, 1917, The-
      odore Roosevelt Digital Library, Dickinson State University, theodorerooseveltcen
      ter.org.

97   "My dear Mr. Leopold": Theodore Roosevelt to Aldo Leopold, January 18, 1917,
      reprinted in Flader, *Thinking Like a Mountain*, xxvii.

98   "with cakes of mud a quarter of an inch thick": Aldo Leopold to Clara Leopold, April
      27, 1920, Aldo Leopold Papers, series 10–8, box 9, folder 2, 166. The trip was to the
      Manzano Mountains and the adjoining Estancia Valley, and Leopold complains of
      snow and "constant wind": "In fact I have concluded the wind *never* blows except in
      the Estancia Valley."

98   Leopold's interest in protecting land as "wilderness": In 1935, Leopold and several
      other prominent conservationists—including Robert Marshall, Robert Sterling
      Yard, and Benton MacKaye—founded the Wilderness Society. For more on the soci-
      ety and the disparate motivations of its founders, see Paul S. Sutter, *Driven Wild:
      How the Fight Against Automobiles Launched the Modern Wilderness Movement* (Seat-
      tle: University of Washington Press, 2005).

98   the highest use was minimal use: For Leopold's ultimate articulation of this idea,
      see "Wilderness," in *Leopold: A Sand County Almanac and Other Writings*, 160–70.

100   "control of other factors": Aldo Leopold, "Progress of the Game Survey," in *Transac-
      tions of the Sixteenth American Game Conference*, December 2–3, 1929, 197–98.

101   his resulting report emphasized the importance of habitat: "Report to the American
      Game Conference on an American Game Policy," in *Transactions of the Seventeenth
      American Game Conference*, December 1–2, 1930, 284–309. For more on the con-
      cept of habitat and its history, see Peter Alagona, "What Is Habitat?" *Environmental
      History* 16 (2011): 433–38.

101   In reaction to exclusionary English game laws: For context, see Mark R. Sigmon,
      "Hunting and Posting on Private Land in America," *Duke Law Journal* 54 (2004):
      549–85.

102   the North American Model of Wildlife Conservation: Valerius Geist et al., "Why
      Hunting Has Defined the North American Model of Wildlife Conservation," *Trans-
      actions of the North American Wildlife and Natural Resources Conference* 66 (2001):
      175–85. For an overview of the model's oversights and limitations, see M. Nils Peter-
      son and Michael Paul Nelson, "Why the North American Model of Wildlife Conser-
      vation Is Problematic for Wildlife Conservation," *Human Dimensions of Wildlife*, no.
      1 (2016): 43–54.

102   a degree of popular fame: Responses to the game survey are described and quoted
      in Meine, *Aldo Leopold*, 278.

102   game survey was published as a book: Aldo Leopold, *Report on the Game Survey of
      the North Central States* (Newtown, CT: Sporting Arms and Ammunition Manufac-
      turers' Institute, 1931).

102   "He insisted that our conquest of nature": Aldo Leopold, "A History of Ideas in

Game Management (from *Game Management*)," in *Leopold: A Sand County Almanac and Other Writings*, 314.

102 **"my whole venture into this field"**: Aldo Leopold to W. T. Hornaday, March 1, 1933, reprinted in *Leopold: A Sand County Almanac and Other Writings*, 779.

103 **"I shall now confide in you"**: Aldo Leopold, "Wherefore Wildlife Ecology?" in *The River of the Mother of God and Other Essays by Aldo Leopold*, edited by Susan L. Flader and J. Baird Callicott (Madison: University of Wisconsin Press, 1991), 336–37.

103 **"In those days, not everybody knew"**: Frances Hamerstrom, *My Double Life: Memoirs of a Naturalist* (Madison: University of Wisconsin Press, 1994), 143.

103 **studies of tropical forests in Panama and elsewhere**: Megan Raby, "An American Tropical Laboratory," in *American Tropics: The Caribbean Roots of Biodiversity Science* (Chapel Hill: University of North Carolina Press, 2017), 21–57.

103 **New York Zoological Society scientist William Beebe**: Carol Grant Gould, *The Remarkable Life of William Beebe* (Washington, DC: Island Press, 2004), 284–93. The bathysphere was invented by engineer Otis Barton.

104 **"daily uncovering a web of interdependencies"**: Aldo Leopold, "Wilderness ('The two great cultural advances . . .')," in *Leopold: A Sand County Almanac and Other Writings*, 375.

104 **"One cannot travel many days in the German forests"**: Quoted in Meine, *Aldo Leopold*, 355–56.

104 **"cabbage brand" forestry**: Aldo Leopold, "A Biotic View of Land," in *Leopold: A Sand County Almanac and Other Writings*, 444.

104 **"cubistic"**: Leopold, "Wilderness ('To an American conservationist . . .')," in *Leopold: A Sand County Almanac and Other Writings*, 372.

104 **some of the movement's leaders drew an explicit connection**: Raymond H. Dominick III, "The Nazis and the Nature Conservationists," *Historian* 49, no. 4 (1997): 508–38.

104 **in turn supported the National Conservation Law**: Frank Uekötter, "Green Nazis? Reassessing the Environmental History of Nazi Germany," *German Studies Review* 30, no. 2 (2007): 275–80.

105 **"a certain exuberance"**: Leopold, "Wilderness (To an American conservationist . . .)," in *Leopold: A Sand County Almanac and Other Writings*, 374.

105 **"first clearly realized that land is an organism"**: Quoted in Flader, *Thinking Like a Mountain*, 153. Also see Aldo Leopold, "Song of the Gavilan," in *Leopold: A Sand County Almanac and Other Writings*, 131–35; Starker Leopold, "Adios, Gavilan," *Pacific Discovery*, January–February 1949, 4–13; and William Fleming and William Forbes, "Following in Leopold's Footsteps: Revisiting and Restoring the Rio Gavilan Watershed," *Ecological Restoration* 24, no. 1 (2006): 25–31.

106 **discarded by many ecologists**: According to Curt Meine, Leopold rejected the "balance of nature" metaphor as early as the 1920s. For a look at its persistence in both ecology and conservation, see John Kricher, *The Balance of Nature: Ecology's Endur-*

ing Myth (Princeton: Princeton University Press, 2009). Also useful is Daniel Simberloff, "The 'Balance of Nature'—Evolution of a Panchreston," *PLoS Biology* 12, no. 10 (2014): e1001963.

106 **"of unprecedented violence, rapidity, and scope"**: Leopold, "A Biotic View of Land," in *Leopold: A Sand County Almanac and Other Writings*, 441.

106 **"the oldest task in human history"**: Aldo Leopold, "Engineering and Conservation," in *Leopold: A Sand County Almanac and Other Writings*, 410.

107 **"a member of a community of interdependent parts"**: Leopold, "The Land Ethic," in *Leopold: A Sand County Almanac and Other Writings*, 172.

107 **"Breakfast comes before ethics"**: Quoted in Meine, *Aldo Leopold*, 504.

107 **"It is going to take patience and money"**: Aldo Leopold, "The Game Situation in the Southwest," Aldo Leopold Papers, series 10–6, box 1, folder 2, 444.

108 **defended "so-called vermin"**: Spiro, *Defending the Master Race*, 84.

108 **"It is a curious perversion"**: Quoted in Peter Alagona, *After the Grizzly: Endangered Species and the Politics of Place in California* (Berkeley: University of California Press, 2013), 78.

108 **"I have never before killed a road-runner"**: Quoted in Alagona, *After the Grizzly*, 78–79.

108 **examined and re-examined**: Margaret Moore et al., "Was Aldo Leopold Right about the Kaibab Deer Herd?," *Ecosystems* 9 (2006): 227–41. Also see Roberta L. Millstein, "Types of Experiments and Causal Process Tracing: What Happened on the Kaibab Plateau in the 1920s," *Studies in History and Philosophy of Science Part A* 78 (2019): 98–104.

109 **"letting the lions alone for a while"**: Quoted in Meine, *Aldo Leopold*, 270.

109 **"reluctantly, selectively"**: Quoted in Meine, *Aldo Leopold*, 299.

109 **"I would infinitely rather that Mr. Burch"**: Aldo Leopold to Karl T. Frederic, December 20, 1935, reprinted in *Leopold: A Sand County Almanac and Other Writings*, 790–91.

109 **"I am in general opposed"**: Aldo Leopold to Rosalie Edge, May 5, 1932, Hawk Mountain Sanctuary Archives.

109 **"those who tell the truth forcefully"**: Rosalie Edge to Aldo Leopold, May 10, 1932, Hawk Mountain Sanctuary Archives.

110 **leveraged the Disney movie *Bambi***: Flader, *Thinking Like a Mountain*, 204. Also see Ralph H. Lutts, "The Trouble with *Bambi*: Walt Disney's *Bambi* and the American Vision of Nature," *Forest and Conservation History* 36, no. 4 (1992): 162.

110 **"the public, unlike the mule"**: Aldo Leopold, "Adventures of a Conservation Commissioner," in *The River of the Mother of God*, 330.

110 **"The lesson you wish to put across"**: Albert Hochbaum to Aldo Leopold, January 22, 1944; Aldo Leopold to Albert Hochbaum, January 29, 1944 ("Your point is obviously well taken"); Albert Hochbaum to Aldo Leopold, February 4, 1944; Albert Hochbaum to Aldo Leopold, March 11, 1944 (marked "Important Letter" in Leo-

pold's handwriting); Aldo Leopold Papers, series 10–6, box 5, folder 1, beginning on pages 263, 260, 256, and 253 respectively. Also see Aldo Leopold to Albert Hochbaum, February 11, 1944; February 12, 1944; and March 1, 1944 ("I think you are partly right, but I am not yet persuaded that you are wholly right"); Aldo Leopold Papers, series 10–2, box 3, folder 6, beginning on pages 903, 905, and 906 respectively. A remarkable exchange.

111 **"In those days, we had never"**: Aldo Leopold, "Thinking Like a Mountain," in *Leopold: A Sand County Almanac and Other Writings*, 114–15.

111 **Just south of the Leopold shack lies the Baraboo Range**: Marcia Bjornerud, *Timefulness* (Princeton: Princeton University Press, 2018), 78.

112 **On April 21, 1948**: Estella B. Leopold, *Stories*, 209–13.

112 **In his pockets**: The contents of Leopold's pockets on the day he died are stored in an envelope in the "Additions to the Collection" series of the Aldo Leopold Papers, 10–13.

113 **a "private discovery"**: Flader, *Thinking Like a Mountain*, ix.

113 **published in a mass-market edition**: Curiously, Leopold's references to evolution were excised from this Ballantine Books edition, and remained absent for many successive printings (the phrase "trumpet in the orchestra of evolution," quoted later in this chapter, is missing from my 1991 copy).

113 **The "only visible remedy"**: Leopold, "The Land Ethic," in *Leopold: A Sand County Almanac and Other Writings*, 180.

113 **"nothing so important as an ethic is ever 'written' "**: Leopold, "The Land Ethic," in *Leopold: A Sand County Almanac and Other Writings*, 188.

113 **"A thing is right when it tends to preserve"**: Leopold, "The Land Ethic," in *Leopold: A Sand County Almanac and Other Writings*, 188.

113 **over the decades his ethic has been variously criticized**: Roberta L. Millstein, "Debunking Myths about Aldo Leopold's Land Ethic," *Biological Conservation* 217 (2018): 391–96.

114 **"a land ethic changes the role"**: Leopold, "The Land Ethic," in *Leopold: A Sand County Almanac and Other Writings*, 173.

115 **"It was a noble sight"**: *Wisconsin Journal*, July 20, 1934, Aldo Leopold Papers, series 10–7, box 2, folder 2, 446–47.

115 **"His tribe, we now know"**: Aldo Leopold, "Marshland Elegy," in *Leopold: A Sand County Almanac and Other Writings*, 86.

115 **Leopold left no clear record**: There is a minor mystery here. According to Stanley Temple, emeritus professor of wildlife ecology at the University of Wisconsin at Madison, the detailed phenology observations that Leopold made at the shack between 1935 and 1945 do not include sandhill cranes, and Estella Jr. wrote in *Stories* that the family's first sighting of cranes at the shack was in the 1970s. But in a remembrance written the day after Leopold's death, Albert Hochbaum recalled "the day at the Shack when [Leopold] showed me my first cranes." See Aldo Leopold Papers, series 10–6, box 5, folder 1, 87.

116 **"The impending industrialization of the world"**: Aldo Leopold, "Post-War Prospects," in *Leopold: A Sand County Almanac and Other Writings*, 492–95. Originally published in *Audubon*, January–February 1944.

CHAPTER FIVE: THE PROFESSOR AND THE ELIXIR OF LIFE

117 **"Dear Grandpater"**: Julian Huxley, *Memories I* (London: Allen and Unwin, 1970), 23.

117 **"They are very wise men"**: Charles Kingsley, *The Water-Babies: A Fairy-Tale for a Land-Baby* (Boston: T. O. H. P. Burnham, 1864), 66.

118 **"My dear Julian"**: Huxley, *Memories I*, 24–25.

119 **"a thing of the past"**: Henry M. Cowles, "A Victorian Extinction: Alfred Newton and the Evolution of Animal Protection," *British Journal for the History of Science* 46, no. 4 (2013): 700.

119 **"a Victorian thinker fated"**: Colin Divall, "From a Victorian to a Modern: Julian Huxley and the English Intellectual Climate," in *Julian Huxley: Biologist and Statesman of Science*, edited by C. Kenneth Waters and Albert Van Helden (College Station: Texas A&M University Press, 1992), 32.

120 **"she drifted hour after hour"**: T. H. Huxley, "Observations upon the Anatomy and Physiology of *Salpa* and *Pyrosoma*," *Philosophical Transactions of the Royal Society of London* 141 (1851): 580.

120 most **"men of science"**: The word "scientist" did not come into common use until the 1900s. For more on Huxley's role in the development of science as a profession, see Ruth Barton, "'Men of Science': Language, Identity and Professionalization in the Mid-Victorian Scientific Community," *History of Science* 41, no. 1 (2003): 73–119.

120 **"his grave, earnest expression"**: Ronald W. Clark, *The Huxleys* (New York: William Heinemann, 1968), 32.

120 **"only afterwards did you realize"**: Quoted in Clark, *The Huxleys*, 79–80.

121 **"seize on each place"**: Charles Darwin, *On the Origin of Species*, reprint of 1859 edition (New York: Signet Classics, 2003), 97.

121 **"I am almost convinced"**: Charles Darwin to Joseph Hooker, January 11, 1844, Darwin Correspondence Project, darwinproject.ac.uk.

121 a **"flash of light"**: T. H. Huxley, "On the Reception of the *Origin of Species*," chapter 1.XIV in *The Life and Letters of Charles Darwin*, vol. 1, edited by Francis Darwin (New York: D. Appleton, 1898), 550.

121 **"How extremely stupid"**: Huxley, "On the Reception of the *Origin of Species*," 551.

121 **"Light will be thrown"**: Darwin, *On the Origin of Species*, 506.

122 **"I trust you will not allow"**: T. H. Huxley to Charles Darwin, November 23, 1859, Darwin Correspondence Project, darwinproject.ac.uk.

122 **"our unsuspected cousinship"**: Samuel Wilberforce, review of *On the Origin of Species*, *Quarterly Review*, 1860, 231. victorianweb.org.

122 **according to the most detailed**: Richard England, "Censoring Huxley and Wilberforce: A New Source for the Meeting That the *Athenaeum* 'Wisely Softened Down,'"
Notes and Records of the Royal Society Journal of the History of Science 71, no. 4 (2017):
371–84.

122 **"If then, said I, the question is put to me"**: T. H. Huxley to Frederick Dyster, September 9, 1860. Quoted in Clark, *The Huxleys*, 59.

122 **"I see daily more & more plainly"**: Charles Darwin to T. H. Huxley, July 20, 1860,
Darwin Correspondence Project, darwinproject.ac.uk.

123 **"The question of questions for mankind"**: T. H. Huxley, *Man's Place in Nature and
Other Anthropological Essays* (New York: D. Appleton, 1919), 77.

123 **"have been exterminated"**: Darwin, *On the Origin of Species*, 351.

124 **"I like that chap!"**: Huxley, *Memories I*, 22.

124 **"bespectacled, bony-faced scholar"**: "The Huxley Brothers," *Life*, March 24, 1947, 53.

124 **"one of the great atheist families"**: Clark, *The Huxleys*, 180.

124 **"I realized more fully than ever"**: Huxley, *Memories I*, 73.

124 **"It is very hard to leave you all"**: Huxley, *Memories I*, 20. Julia Huxley was the niece
of the poet and critic Matthew Arnold, and her farewell quoted his poem "Rugby
Chapel": "I am only going into another room 'of the sounding labour-house vast of
Being.'"

126 **"a gallant little machine"**: Huxley, *Memories I*, 100.

126 **"all the familiar institutions"**: Clark, *The Huxleys*, 173.

126 **"which are either useful to man or are harmless"**: Opening lines of the Convention
for the Preservation of Wild Animals, Birds, and Fish in Africa, adopted in London
in 1900, text republished by the International Environmental Agreements Database Project, iea.uoregon.edu. For context, see Rachelle Adam, *Elephant Treaties:
The Colonial Legacy of the Biodiversity Crisis* (Lebanon, NH: University Press of New
England, 2014), 23–25.

127 **a group of "penitent butchers"**: The group embraced the nickname to the point that
its organizational memoir, published in 1978, is titled *The Penitent Butchers*.

127 **The first such treaty**: Robert Boardman, *International Organization and the Conservation of Nature* (London: Macmillan, 1981), 27–28.

127 **Swiss conservationist Paul Sarasin**: For more about Sarasin, see Jan-Henrik Meyer,
"From Nature to Environment: International Organizations and Environmental
Protection Before Stockholm," in *International Organizations and Environmental
Protection: Conservation and Globalization in the Twentieth Century*, edited by Wol-
fram Kaiser and Jan-Henrik Meyer (New York: Berghahn, 2017), 31–73.

127 **"All the wild, living higher fauna of our planet"**: Boardman, *International Organization*, 29.

128 **"He lacked the pensive beauty of Aldous"**: Juliette Huxley, *Leaves of the Tulip Tree:
Autobiography* (London: John Murray, 1986), 57. Dolichocephalic, meaning "long-
headed," was a term coined by anthropologists in the 1850s to describe one of the

three primary categories of human "cephalic indices" or skull shapes. Initially used to classify prehistoric human remains, the categories were exploited by "race scientists" such as Madison Grant, who used skull shape and other physical characteristics to argue for the existence of a racial hierarchy and the superiority of the "Nordic race."

128 **"impatient of intrusion or contradiction"**: Huxley, *Leaves*, 76.

128 **beginning a passionate correspondence**: *Dear Juliette: Letters of May Sarton to Juliette Huxley*, edited by Susan Sherman (New York: W. W. Norton, 1999).

128 **"Julian had the gift of seeing"**: Huxley, *Leaves*, 74.

128 **reported his results in a letter to *Nature***: Julian Huxley, "Metamorphosis of Axolotl Caused by Thyroid-Feeding," *Nature* 104 (1920): 435.

128 **discovered the "Elixir of Life"**: This was a headline in the *Pall Mall Gazette*; "one of the most brilliant of our young biologists" was from the *Daily Mail*, which broke the story of Huxley's study. Both are quoted in Jan Witowski, "The Elixir of Life," *Trends in Genetics* 2 (1986): 110–13.

130 **"For goodness sake do decide"**: Quoted in Steindór J. Erlingsson, "The Costs of Being a Restless Intellect: Julian Huxley's Popular and Scientific Career in the 1920s," *Studies in History and Philosophy of Biological and Biomedical Sciences* 40 (2009): 104.

130 **"the centre and focus of popular interest"**: Quoted in Joe Cain, "Julian Huxley, General Biology, and the London Zoo, 1935–42," *Notes and Records of the Royal Society* 64 (2010): 368.

130 **"the only horseman ever unseated in England"**: Clark, *The Huxleys*, 258.

130 **"He's a biter, not a kicker"**: Quoted in Clark, *The Huxleys*, 266.

130 **"the display of our Menagerie"**: Quoted in Cain, "Julian Huxley and the London Zoo," 370.

131 **Huxley named Don Quixote**: "The BBC Brains Trust Answering 'Any Questions?'" reel 3, 1945. British Pathé YouTube channel.

131 **"I seem to have been possessed"**: Huxley, *Memories I*, 7.

131 **"no hope that mere human beings"**: T. H. Huxley, "Prolegomena," in *Evolution and Ethics and Other Essays* (New York: D. Appleton, 1899), 34.

131 **British philosopher Herbert Spencer**: Spencer first used the phrase "survival of the fittest" in his *Principles of Biology*, published in 1864. Darwin, in the 1869 edition of *On the Origin of Species*, adopted it as a synonym for natural selection, but defined "fittest" more clearly and narrowly than Spencer had.

131 **"constant tendency in all animated life"**: Thomas Malthus, *An Essay on the Principle of Population*, vol. 1 (Washington, DC: Weightman, 1809), 7.

132 **"worst fitted"**: Herbert Spencer, *Social Statics* (New York: D. Appleton, 1873), 416.

132 **"ultimate perfection"**: Spencer, *Social Statics*, 353.

132 **"poverty of the incapable"**: Spencer, *Social Statics*, 354.

132   **neither finite nor destined to improve humanity**: Michael Ruse, "Social Darwinism: The Two Sources," *Albion* 12, no. 1 (1980): 23–36.

132   **"a drunkard's walk"**: Stephen Jay Gould, *Life's Grandeur: The Spread of Excellence from Plato to Darwin* (London: Jonathan Cape, 1996), 149. The degree to which evolution is or is not "progressive" remains a matter of some debate. See Timothy Shanahan, "Evolutionary Progress?," *BioScience* 50, no. 5 (2000): 451–59.

132   **"social Darwinism"**: Darwin's eventual acceptance of some of the principles of "social Darwinism" is traced by Gregory Claeys in "The 'Survival of the Fittest' and the Origins of Social Darwinism," *Journal of the History of Ideas* 61, no. 2 (2000): 223–40.

132   **"the most highly developed and perfect"**: Ernst Haeckel, *The History of Creation*, vol. 2, translated by E. Ray Lankester (London: Kegan Paul, Trench, Trübner, 1899), 429. See also Richard Weikart, "The Origins of Social Darwinism in Germany, 1859–1895," *Journal of the History of Ideas* 54, no. 3 (1993): 469–88.

132   **Darwin's cousin Francis Galton**: For more about Galton, including his testy relationship with Darwin, see Nicholas Gillham, "Sir Francis Galton and the Birth of Eugenics," *Annual Review of Genetics* 35 (2001): 83–101.

132   **"What nature does blindly"**: Francis Galton, "Eugenics: Its Definition, Scope and Aims," *American Journal of Sociology* 10, no. 1 (1904): 1–25.

133   **best suited to a situation**: Herbert Spencer, "Evolutionary Ethics," in *Various Fragments* (New York: D. Appleton, 1898), 125. While Spencer affirmed here that "the survival of the fittest is not always the survival of the best," T. H. Huxley believed that Spencer's choice of language, and Darwin's acceptance of it, had confused the public's understanding of evolution. "The unlucky substitution of 'survival of fittest' for 'natural selection' has done much harm in consequence of the ambiguity of 'fittest'—which many take to mean 'best' or 'highest'—whereas natural selection may work towards degradation," he wrote (T. H. Huxley to W. Platt Ball, October 27, 1890).

133   **"I wish very much"**: Theodore Roosevelt, "Twisted Eugenics," *Outlook* 106 (January 1914): 32. Roosevelt expressed this in the course of an argument for "positive" rather than punitive eugenic policies: "The emphasis should be laid on getting desirable people to breed."

133   **"nothing but applied biology"**: Recollection of an SS doctor interviewed by Robert Jay Lifton for *The Nazi Doctors: Medical Killing and the Psychology of Genocide* (New York: Basic Books, 1986), 31.

133   **influential United Nations statements**: See *Four Statements on the Race Question* (Paris: UNESCO, 1969). Huxley contributed to two of these statements, which became more cautious over time. See Michelle Brattain, "Race, Racism, and Anti-Racism: UNESCO and the Politics of Presenting Science to the Public," *American Historical Review* 112, no. 5 (2007): 1386–1413.

133   **a leading member of the British Eugenics Society**: Paul Weindling, "Julian Huxley

and the Continuity of Eugenics in Twentieth-Century Britain," *Journal of Modern European History* 10, no. 4 (2012): 480–99.

133   **laws adopted by more than thirty U.S. states:** Alexandra Minne Stein, *Eugenic Nation: Faults and Frontiers of Better Breeding in Modern America* (Oakland: University of California Press, 2016), 115.

133   **"single equalized environment":** Garland E. Allen, "Julian Huxley and the Eugenical View of Human Evolution," in *Julian Huxley: Biologist and Statesman of Science* (College Station: Texas A&M University Press, 1992).

133   **"unteachables":** Allen, "Julian Huxley and the Eugenical View of Human Evolution," 212.

134   *If I Were Dictator:* Julian Huxley, *If I Were Dictator* (New York: Harper and Brothers, 1934).

134   **"He whirls indefatigably":** Quoted in Stephen Spender, "The Conscience of the Huxleys," *New York Review of Books*, March 25, 1971.

134   **"I repeat again":** Quoted in Gifford Pinchot, *Breaking New Ground* (Washington, DC: Island Press, 1998), 371.

135   **the war's cascading disasters:** Thomas Robertson, "Total War and the Total Environment: Fairfield Osborn, William Vogt, and the Birth of Global Ecology," *Environmental History* 17 (2012): 336–64.

135   **"The antipathies of nations and races":** Fairfield Osborn, *Our Plundered Planet* (New York: Little, Brown, 1970), 32.

135   **reports of environmental ruin:** See Frank Graham Jr., *The Audubon Ark: A History of the National Audubon Society* (Austin: University of Texas Press, 1992), 142–44.

136   **he packed a copy of *Game Management*:** Gregory T. Cushman, "'The Most Valuable Birds in the World': International Conservation Science and the Revival of Peru's Guano Industry, 1909–1965," *Environmental History* 10, no. 3 (2005): 494.

136   **"The guano bird population has been withdrawing":** William Vogt to Aldo Leopold, July 29, 1939, Aldo Leopold Papers, University of Wisconsin Archives, series 10–1, box 3, folder 8, 362.

136   **"pitiful, collapsed, downy clumps":** William Vogt, *People! Challenge to Survival* (New York: William Sloane, 1960), 125.

136   **"land sickness":** Quoted in Gregory T. Cushman, *Guano and the Opening of the Pacific World: A Global Ecological History* (Cambridge: Cambridge University Press, 2013), 256–57.

136   **"An eroding hillside in Mexico":** William Vogt, *Road to Survival* (New York: William Sloan, 1948), 285.

136   **"spawning millions":** Vogt, *Road to Survival*, 236.

136   **"recklessly and irresponsibly":** Vogt, *Road to Survival*, 76.

137   **"responsible for more millions":** Vogt, *Road to Survival*, 48.

137   **"not only desirable":** Vogt, *Road to Survival*, 238.

137   **the ecological concept of carrying capacity:** Nathan F. Sayre, "The Genesis, History,

and Limits of Carrying Capacity," *Annals of the Association of American Geographers* 98, no. 1 (2008): 120–34. Also see David Price, "Carrying Capacity Reconsidered," *Population and Environment* 21, no. 1 (1999), 5–26.

137   **"dead of its own too-much"**: Aldo Leopold, "Thinking Like a Mountain," in *Leopold: A Sand County Almanac and Other Writings on Ecology and Conservation*, edited by Curt Meine (New York: Library of America, 2013), 116. This phrase echoes Shakespeare's warning, in *Hamlet*, that "goodness, growing to a pleurisy / Dies in his own too-much," which Leopold references in "Wilderness (To an American conservationist . . .)," in *Leopold: A Sand County Almanac and Other Writings*, 374.

137   **For Leopold, it was the land ethic**: Leopold had at least some sympathy for Vogt's position on food aid, though his criticism was far milder and never published. See Miles A. Powell, *Vanishing America: Species Extinction, Racial Peril, and the Origins of Conservation* (Cambridge, MA: Harvard University Press, 2016), 174.

138   **a 1965 essay for *Playboy***: Julian Huxley, "The Age of Overbreed," *Playboy*, January 1965, 103–4.

138   **the guiding purpose of UNESCO**: For more on the relationship between Huxley's personal philosophy and UNESCO's mission, see Glenda Sluga, "UNESCO and the (One) World of Julian Huxley," *Journal of World History* 21, no. 3 (2010): 393–418.

138   **"Delegates asked what seemed"**: Julian Huxley, *Memories II* (London: Allen and Unwin 1973), 50–51.

139   **"collect, analyse, interpret"**: Quoted in Martin Holdgate, *The Green Web* (New York: Earthscan, 1999), 32.

139   **"the fascination of all these other manifestations"**: Quoted in Holdgate, *The Green Web*, 33.

139   **"a data gap of awesome proportions"**: Boardman, *International Organization*, 47.

139   **"scattered in many minds and documents"**: Leopold, "Threatened Species: A Proposal to the Wildlife Conference for an Inventory of the Needs of Near-Extinct Birds and Animals," in *Leopold: A Sand County Almanac and Other Writings*, 378.

139   **"the entire world biotic community"**: Quoted in Stephen J. Macekura, *Of Limits and Growth: The Rise of Global Sustainable Development in the Twentieth Century* (Cambridge: Cambridge University Press, 2015), 35.

140   **Eight of those twenty-seven species**: The other species on the list that are now believed extinct are the thylacine, the carnivorous Australian marsupial sometimes called the Tasmanian tiger; the Arabian ostrich (a subspecies); the bubal hartebeest (a subspecies); the Caribbean monk seal; the Cuban ivory-billed woodpecker (a subspecies); the Eskimo curlew; the Marinara mallard (sometimes considered a subspecies); and the pink-headed duck. See Richard Fitter, "25 Years On: A Look at Endangered Species," *Oryx* 12, no. 3 (1974): 341–46.

140   **His report, *A Look at Threatened Species***: Lee Talbot, "A Look at Threatened Species: A Report on Some Animals of the Middle East Which Are Threatened with Extermination," *Oryx* 5, no. 4–5 (1960): 155–293.

140 **the soon-to-be iconic Red Data Books**: Peter Scott et al., "Red Data Books: The Historical Background," in *The Road to Extinction: Proceedings of a Symposium Held by the Species Survival Commission*, edited by Richard and Maisie Fitter (Gland, Switzerland: IUCN, 1987), 1–6.

141 **"On the beautiful shores of the Bosphorus"**: Quoted in Huxley, *Memories II*, 66.

141 **Juliette accompanied him**: Juliette Huxley published a memoir of the trip called *Wild Lives of Africa* (London: Collins, 1963).

141 **a vivid three-part series**: Julian Huxley, "The Treasure House of Wild Life," *Observer*, November 13, 1960, 23–24; "Cropping the Wild Protein," *Observer*, November 20, 1960, 23; "Wild Life as a World Asset," *Observer*, November 27, 1960, 23.

142 **"There must be a way to the conscience"**: Quoted in Alex Schwarzenbach, *Saving the World's Wildlife: WWF—The First 50 Years* (London: Profile, 2011), 15–18. While Nicholson adopted Stolan's suggestion, he quickly dismissed the man himself as a "naive enthusiast."

143 **a Yellowstone National Park campfire**: Richard Fitter and Peter Scott, *The Penitent Butchers: 75 Years of Wildlife Conservation* (Cambridge: Fauna Preservation Society, 1978), 16–17.

143 **"The question African people ask"**: Malumo P. Simbotwe, "African Realities and Western Expectations," in *Voices from Africa: Local Perspectives on Conservation*, edited by Dale Lewis and Nick Carter (Washington, DC: World Wildlife Fund, 1993), 15.

144 **"conservation of wild life and wild places"**: Quoted in Macekura, *Of Limits and Growth*, 57–61.

144 **"not even natives"**: Quoted in Macekura, *Of Limits and Growth*, 62.

144 **slaughtering rhinos in protest**: David Western, *In the Dust of Kilimanjaro* (Washington, DC: Island Press, 2001), 128.

145 **"it was one of my most pleasant trips"**: Terry Spencer to Stewart Udall, November 26, 1963, Stewart L. Udall Papers (AZ 372), Special Collections, University of Arizona Libraries, box 111, folder 2.

146 **"I have unveiled these sorry episodes"**: "Address by Secretary of the Interior Stewart L. Udall at Eighth General Assembly of the International Union for Conservation of Nature and Natural Resources, Nairobi, Kenya, September 16, 1963," U.S. Department of the Interior press release, Stewart L. Udall Papers (AZ 372), Special Collections, University of Arizona Libraries, box 111, folder 1.

CHAPTER SIX: THE EAGLE AND THE WHOOPING CRANE

148 **"They came by like brown leaves"**: Rachel Carson, "Road of the Hawks," in *Lost Woods: The Discovered Writing of Rachel Carson*, edited by Linda Lear (Boston: Beacon Press, 1998), 30–32.

150 **"As you may possibly have heard"**: Rosalie Edge to Maurice Broun, May 13, 1960.

Reproduced in Dyana Z. Furmansky, *Rosalie Edge, Hawk of Mercy: The Activist Who Saved Nature from the Conservationists* (Athens, GA: University of Georgia Press, 2009), in the photo insert preceding chapter 7.

151 **"We have discovered many preventives"**: Winston Churchill, speech to the House of Commons, September 28, 1944. *Hansard* 403, column 480.

151 **"Neapolitans are now throwing DDT at brides"**: "The Conquest of Typhus," *New York Times*, June 4, 1944, 88. Presciently, the article notes that "DDT seems almost too good to be true."

151 **Photos from the time**: "Typhus in Naples," *Life*, February 28, 1944, 36–37.

151 **"the War's greatest contribution"**: James Stevens Simmons, "How Magic Is DDT?" *Saturday Evening Post*, January 6, 1945, 86.

152 **"more than ordinary interest and importance"**: Quoted in Linda Lear, *Rachel Carson: Witness for Nature* (Boston: Mariner, 1997), 118–19.

152 **"nineteen robins were found dead"**: Quoted in Furmansky, *Rosalie Edge*, 244.

152 **A study in Michigan suggested**: Roy J. Barker, "Notes on Some Ecological Effects of DDT Sprayed on Elms," *Journal of Wildlife Management* 22, no. 3 (1958): 269–74.

153 **"recognized and admitted by the defendants"**: Quoted in Harold Schmeck Jr., "Long Islanders Ask Court to Halt DDT War on Moth as Health Risk," *New York Times*, May 9, 1957, 1.

153 **"presented no evidence"**: *Murphy v. Benson*, 164 F. Supp. 120 (E.D.N.Y. 1958).

153 **which voted not to review the case**: *Murphy v. Butler*, 362 U.S. 929 (1960). Justice William O. Douglas dissented from the court's decision not to review.

153 **plaintiff Marjorie Spock**: John Paull, "The Rachel Carson Letters and the Making of *Silent Spring*," *SAGE Open* 3, no. 3 (2013): 1–12.

153 **"especially significant"**: Rachel Carson, *Silent Spring* (Boston: Houghton Mifflin, 1987), 119. Originally published in 1962.

154 **a three-part series in *The New Yorker***: The series began with "Silent Spring—1," on June 16, 1962, and continued on June 23 and June 30.

154 **the physician and anti-nuclear activist Albert Schweitzer**: Schweitzer was also known for the ethical philosophy he called "reverence for life"—"good consists in maintaining, assisting and enhancing life, and to destroy, to harm or to hinder life is evil"—a worldview shared by many in the environmental movement.

154 **"emotional and inaccurate outburst"**: "Pesticides: The Price for Progress," *Time*, September 18, 1962.

154 **"Silence, Miss Carson!"**: Quoted in Michael B. Smith, "'Silence, Miss Carson!' Science, Gender, and the Reception of Silent Spring," *Feminist Studies* 27, no. 3 (2001): 733–52.

154 **"sparing, selective and intelligent use"**: Quoted in Jonathan Norton Leonard, "Rachel Carson Dies of Cancer; 'Silent Spring' Author Was 56," *New York Times*, April 15, 1964.

154 **"Who has decided"**: Carson, *Silent Spring*, 127.

155   **"play the same role as the wolves"**: Carson, *Silent Spring*, 248.

155   **"We are losing half the subject-matter"**: As recalled by Julian Huxley in his preface to the UK edition of *Silent Spring* (London: Penguin, 1991), 20.

155   **"freed themselves from the fears and superstitions"**: Rachel Carson, "The Pollution of Our Environment," in *Lost Woods*, 244–45.

155   **he kept Carson's books**: Douglas Brinkley, "Rachel Carson and JFK, an Environmental Tag Team," *Audubon*, May–June 2012.

155   **to Udall's chagrin**: Stewart Udall, transcript of recorded interview by W. W. Moss, February 16, 1970, John F. Kennedy Library Oral History Program, 35–37.

156   **a sense of obligation to the land**: Barbara Leunes, "The Conservation Philosophy of Stewart L. Udall, 1961–1968," PhD diss., Texas A&M University, 1977, 19–26.

156   **a "vigorous program"**: Stewart Udall, transcript of recorded interview #2 by Joe B. Frantz, May 19, 1969, Lyndon Baines Johnson Library Oral History Collection, 6.

156   **"Rachel was a friend"**: Stewart Udall, speech at the Rachel Carson Centennial, June 2, 2007, John F. Kennedy Presidential Library, transcript and video at cspan.org.

156   **"awakened the Nation"**: Quoted in William Souder, *On a Farther Shore: The Life and Legacy of Rachel Carson* (New York: Penguin Random House, 2012), 379–80.

156   **"During the years I worked on *Silent Spring*"**: Rachel Carson to Stewart Udall, May 3, 1963, Stewart L. Udall Papers (AZ 372), Special Collections, University of Arizona Libraries, box 190, folder 5.

156   **"including, of course, Dreamcat"**: Rachel Carson to Stewart Udall, May 13, 1963, Stewart L. Udall Papers (AZ 372), Special Collections, University of Arizona Libraries, box 190, folder 5.

157   **"Somehow the sharing of beautiful and lovely things"**: Rachel Carson to Dorothy Freeman, June 12, 1955, in *Always, Rachel: The Letters of Rachel Carson and Dorothy Freeman, 1952–1964*, edited by Martha Freeman (Boston: Beacon Press, 1995), 155.

158   **"Secretary of Things in General"**: Robert Manning, "Secretary of Things in General," *Saturday Evening Post*, May 20, 1961, 38–39.

158   **demanding . . . that the Washington, DC, football franchise integrate its team**: The team, which at the time was owned by George Preston Marshall, was the last in major league sports to integrate. Not until 2020, after years of public pressure, did the team's current ownership agree to drop the name Marshall gave it in 1933: the Redskins. For more on both battles and their intersection, see Joel Anderson, Stefan Fatsis, and Josh Levin, "Will Washington's NFL Team Finally Change Its Racist Nickname?," *Hang Up and Listen*, podcast, June 22, 2020.

158   **"I have no doubt"**: Quoted in Leunes, "The Conservation Philosophy of Stewart L. Udall," 15.

159   **"Contemplating the teeming life of the shore"**: Rachel Carson, *The Edge of the Sea* (Boston: Houghton Mifflin, 1998), 250. Originally published in 1955.

159   **the word "environment" was usually a synonym**: Paul Warde et al., *The Environment: A History of the Idea* (Baltimore: Johns Hopkins University Press, 2018), 7–8.

159 **Early environmentalists brought middle-class attention**: Robert Gottlieb, "Urban and Industrial Roots," chapter 2 in *Forcing the Spring: The Transformation of the American Environmental Movement* (Washington, DC: Island Press, 2005), 83–120.

160 **she responded with a blistering letter**: Rachel Carson to Fon Boardman of Oxford University Press, circa September 1953. I am indebted to Curt Meine for sharing an image of the original letter and its marginal note. The text of the letter is reprinted in *Rachel Carson: Silent Spring and Other Writings on the Environment*, edited by Sandra Steingraber (New York: Library of America, 2018), 327.

160 **"land ethic for tomorrow"**: Stewart L. Udall, *The Quiet Crisis* (New York: Avon, 1963), 202.

160 **"total American environment"**: Stewart Udall, transcript of recorded interview #2 by Joe B. Frantz, Lyndon Baines Johnson Library Oral History Collection, 20.

160 **"higher order of conservation statesmanship"**: Quoted in Leunes, "The Conservation Philosophy of Stewart L. Udall," 62.

160 **"Widening concept of conservation"**: Stewart Udall, draft address to the IUCN, Stewart L. Udall Papers (AZ 372), Special Collections, University of Arizona Libraries, box 111, folder 1, 12–13.

161 **the *New York Times Magazine* published**: Stewart Udall, "To Save Wildlife and Aid Us, Too," *New York Times Magazine*, September 15, 1963, 44–45.

161 **"We have cranes of two colors"**: Micah True, *The Jesuit Pierre-François-Xavier de Charlevoix's (1682–1761) Journal of a Voyage in North America: An Annotated Translation* (Leiden: Brill, 2019), 194. For clarity I substituted "grayish" for "gridelin" or *gris-de-lin*, a reddish or purplish gray.

161 **"the most part white"**: Quoted in Robert Porter Allen, *The Whooping Crane*, National Audubon Society Research Report No. 3, June 1952, 5.

162 **"numerous legions"**: Quoted in Allen, *The Whooping Crane*, 11.

162 **"enthusiastic imagination"**: Allen, *The Whooping Crane*, 10.

162 **"though common in Texas and Florida"**: Quoted in Allen, *The Whooping Crane*, 14.

162 **"this splendid bird will almost certainly"**: Quoted in Allen, *The Whooping Crane*, 70.

162 **"beyond saving"**: Waldron DeWitt Miller et al., "A Crisis in Conservation," Hawk Mountain Sanctuary Archives, 2.

164 **"I wondered idly what poor, unsuspecting soul"**: Robert Porter Allen, *On the Trail of Vanishing Birds* (New York: McGraw-Hill, 1957), 37.

164 **"We had gazed at its features"**: Allen, *On the Trail of Vanishing Birds*, 231.

165 **"Now we will be shot, stuffed"**: George Archibald, *My Life with Cranes: A Collection of Stories* (Baraboo, WI: International Crane Foundation, 2016), 45.

165 **"And we ain't talking about"**: Quoted in Thomas R. Dunlap, "Organization and Wildlife Preservation: The Case of the Whooping Crane in North America," *Social Studies of Science* 21, no. 2 (1991): 206–7.

166 **"backlog of information"**: Minutes of Whooping Crane Conference, October 29,

1956, Smithsonian Institution Archives, Accession T89021, National Fish and
Wildlife Laboratory, Bird Project Records, box 1, 9.

166 **"Everybody was at each other's throats"**: Ray Erickson, transcript of recorded interview with Erickson and Dave Marshall by Donna Stoball and Mark Madison, September 12, 2006, U.S. Fish and Wildlife Service, National Conservation Training
Center Museum and Archives, 14.

166 **once writing to a colleague**: Dunlap, "Organization and Wildlife Preservation,"
204.

166 **"Is scarcity to be our only standard of value"**: Stewart Udall, "What Price Resources
for the Good Life?," in *Transactions of the Twenty-Ninth North American Wildlife and
Natural Resources Conference*, March 9–11, 1964, 57.

167 **"pursue, hunt, take"**: Migratory Bird Treaty Act of 1918, 16 U.S.C. § 703–12.

167 **"official designation of rare and endangered"**: U.S. Department of the Interior
Committee Information Sheet, 1964, Smithsonian Institution Archives, Accession
T89021, National Fish and Wildlife Laboratory, Bird Project Records, box 7, 1.

168 **While some prominent ecologists had begun to study entire ecosystems**: Two of
the best-known were the brothers Eugene and H. T. Odum, who carried out their
early research on coral reef metabolism in the wake of atomic bomb tests on the
Pacific atoll of Eniwetok. See Eugene P. Odum, "The New Ecology," *Ecology* 14, no.
7 (1964): 14–16, and "The Emergence of Ecology as a New Integrative Discipline,"
Science 195 (1977): 1289–93.

168 **"significantly constrained the shape"**: Johnny Winston, "Science, Practice, and Policy: The Committee on Rare and Endangered Wildlife Species and the Development of U.S. Federal Endangered Species Policy, 1956–1973," PhD diss., Arizona
State University, 2011, 54.

168 **"conserve, protect, restore, and where necessary"**: Quoted in Steven Lewis Yaffee,
Prohibitive Policy: Implementing the Federal Endangered Species Act (Cambridge,
MA: MIT Press, 1982), 38–39.

168 **"insofar as is practicable"**: Quoted in Yaffee, *Prohibitive Policy*, 40–41.

169 **"ecosystems upon which endangered species"**: Quoted in Winston, "Science, Practice, and Policy," 156.

170 **One analysis of a subset of species recovery plans**: Delisting-date data is from
Kieran Suckling et al., "On Time, On Target: How the Endangered Species Act is
Saving America's Wildlife," Center for Biological Diversity, 2012, 5–6.

171 **Wurster, as he remembered it**: Charles F. Wurster, *DDT Wars: Rescuing Our
National Bird, Preventing Cancer, and Creating the Environmental Defense Fund*
(Oxford: Oxford University Press, 2015), 17.

172 **a "single, interrelated system"**: Richard Nixon, "Message of the President," in *Reorganization Plan No. 3 of 1970*, 35 Fed. Reg. 15623 (July 9, 1970).

172 **"I am persuaded"**: Environmental Protection Agency, "Consolidated DDT Hear-

ings: Opinion and Order of the Administrator" (William F. Ruckelshaus), *Federal Register* 37, no. 131 (July 7, 1972), 13369.

176 **By the end of 2019, forty-two parent-reared chicks**: Hillary L. Thompson et al., "Effects of Release Techniques on Parent-Reared Whooping Cranes in the Eastern Migratory Population," manuscript submitted for publication.

176 **"This is a bird that cannot compromise"**: Allen, *The Whooping Crane*, 14.

176 **"We have singled out the Whooping Crane"**: Allen, *The Whooping Crane*, 204.

178 **In a 1971 essay**: Stewart L. Udall, "The Ivory Tower Is Under Siege," *Audubon*, March 1971, 118–19. The essay was based on a speech given by Udall at the 137th annual meeting of the American Association for the Advancement of Science, held in Chicago in December 1970.

178 **"Qualified biologists are beginning to forge"**: David Ehrenfeld, *Biological Conservation* (New York: Holt, Reinhart, and Winston, 1970), vii.

CHAPTER SEVEN: THE SCIENTISTS WHO ESCAPED THE TOWER

179 **Michael Soulé addressed an audience**: The meeting is described in Ann Gibbons, "Conservation Biology in the Fast Lane," *Science* 255 (1992): 20–23.

179 **a "crisis discipline"**: Michael Soulé, "What Is Conservation Biology?," in *Collected Papers of Michael E. Soulé: Early Years in Modern Conservation Biology* (Washington, DC: Island Press, 2014), 31.

180 **the "ecology movement"**: See David Sills, "The Environmental Movement and its Critics," *Human Ecology* 3, no. 1 (1975): 1–41.

180 **"Disciplines are not logical constructs"**: Soulé, "Conservation Biology and the 'Real World,'" in *Conservation Biology: The Science of Scarcity and Diversity*, edited by Michael Soulé (Sunderland, MA: Sinauer Associates, 1986), 3.

180 **"And think, also, what it would mean"**: Quoted in Robert A. Croker, *Pioneer Ecologist: The Life and Work of Victor Ernest Shelford, 1877–1968* (Washington, DC: Smithsonian Institution Press, 1991), 122.

181 **"meddling with Divinity"**: Quoted in Charles Weld, *A History of the Royal Society: With Memoirs of the Presidents*, vol. 1 (London: John Parker, 1848), 146.

181 **"the extinction of the numerous forms of life"**: Quoted in Mark V. Lomolino, "Wallace, Biogeography, and Conservation Biology," in *An Alfred Russel Wallace Companion*, edited by Charles H. Smith et al. (Chicago: University of Chicago Press, 2019), 350–51.

181 **"It is my hope"**: Francis Sumner, "The Responsibility of the Biologist in Preserving Natural Conditions," *Science* 54 (1921): 39–43.

182 **"Human society, which supports research"**: Victor Shelford, "The Conflict Between Science and Biological Industry," *Science* 100 (1944): 451.

182 **the protection of "biotic communities"**: "Permanent Constitution: Ecologists

Union," December 31, 1947, Aldo Leopold Papers, University of Wisconsin Archives, series 10–2, box 2, folder 13, 1050.

183 **"When I'm in a place with many creatures"**: Quoted in Leath Tonino, "We Only Protect What We Love: Michael Soulé on the Vanishing Wilderness," interview, *Sun Magazine*, April 2018.

184 **lizards on larger islands closer to the mainland**: Michael Soulé, "Phenetics of Natural Populations. III. Variation in Insular Populations of a Lizard," *American Naturalist* 106, no. 950 (1972): 429–46.

184 **Using an early method of genetic analysis**: Michael Soulé, "Island Lizards: The Genetic–Phenetic Variation Correlation," *Nature* 242 (1973): 191–93.

184 **MacArthur and Wilson proposed**: Robert H. MacArthur and Edward O. Wilson, "An Equilibrium Theory of Insular Zoogeography," *Evolution* 17, no. 4 (1963): 373–87.

184 **Wilson and . . . Simberloff tested this theory**: See Daniel Simberloff and Edward Wilson, "A Pioneering Adventure Becomes an Ecological Classic: The Pioneers," *Bulletin of the Ecological Society of America* 98, no. 4 (2017): 276–77.

185 **"The streets seemed alive with people"**: Paul Ehrlich, *The Population Bomb* (New York: Ballantine, 1968), 15.

186 **"It's really very simple, Johnny"**: Quoted in Paul Sabin, *The Bet: Paul Ehrlich, Julian Simon, and Our Gamble Over Earth's Future* (New Haven: Yale University Press, 2013), 2.

187 **"How about it, doomsayers"**: Julian L. Simon, "Environmental Disruption or Environmental Improvement?," *Social Science Quarterly* 62, no. 1 (1981): 39.

187 **the concept of "planetary boundaries"**: Johan Rockström et al., "A Safe Operating Space for Humanity," *Nature* 461 (2009): 472–75.

188 **"There is no reason to expect"**: Kingsley Davis, "Population Policy: Will Current Programs Succeed?," *Science* 158 (1967): 732.

188 **"people *want* large families"**: Ehrlich, *Population Bomb*, 83.

188 **"preventive checks"**: Thomas Malthus, *An Essay on the Principle of Population*, vol. 1 (Washington, DC: Weightman, 1809), 579.

188 **"moral restraint"**: Malthus, *An Essay on the Principle of Population*, 19.

189 **Countries including Bangladesh, Kenya, Indonesia, and Iran**: John Bongaarts and Stephen Sinding, "Population Policy in Transition in the Developing World," *Science* 33 (2011): 574–76.

189 **"When wisdom dictates"**: Neil MacFarquhar, "With Iran Population Boom, Vasectomy Receives Blessing," *New York Times*, September 8, 1996, 1.

189 **"women's ability to control"**: "Programme of Action," International Conference on Population and Development, September 5–13, 1994. Twentieth anniversary edition, 26.

189 **essentially level off by the year 2100**: United Nations Department of Economic and Social Affairs, "World Population Prospects 2019." Also see Anthony Cilluffo and

Neil Ruiz, "World's Population Is Projected to Nearly Stop Growing by the End of the Century," Pew Research Center, June 17, 2019.

190 **an update to the planetary boundaries framework**: Will Steffen et al., "Planetary Boundaries: Guiding Human Development on a Changing Planet," *Science* 347 (2015): 1259855.

190 **his *Science* essay "The Tragedy of the Commons"**: Garrett Hardin, "The Tragedy of the Commons," *Science* 162 (1968): 1243–48.

190 **"The less provident and less able"**: Garrett Hardin, "Lifeboat Ethics: The Case Against Helping the Poor," *Psychology Today*, September 1974. Republished by garretthardinsociety.org.

190 **what Hardin called a tragedy is really a drama**: National Research Council, *The Drama of the Commons*, edited by Elinor Ostrom et al. (Washington, DC: National Academy Press, 2002). For a more recent takedown of Hardin's thesis, and an examination of the untapped potential of the commons concept, see Catherine Brinkley, "Hardin's Imagined Tragedy Is Pig Shit: A Call for Planning to Recenter the Commons," *Planning Theory* 19, no. 1 (2020), 127–44.

191 **"treasuries of variation"**: Otto Frankel and Erna Bennett, "Genetic Resources: Introduction," in *Genetic Resources in Plants: Their Exploration and Conservation*, edited by Otto Frankel and Erna Bennett (Oxford: Blackwell Scientific Publications, 1970), 9.

193 **called "The Biological Diversity Crisis"**: Edward O. Wilson, "The Biological Diversity Crisis," *BioScience* 35, no. 11 (1985): 700–06.

194 **the National Forum on BioDiversity**: E. O. Wilson, ed., *Biodiversity* (Washington, DC: National Academies Press, 1988), v.

194 **Wilson told the virtual audience**: Quotes are from "Biodiversity," a series of excerpts from the 1986 National Teleconference on Biodiversity, vimeo.com/academies.

195 **A search of one comprehensive index**: Timothy J. Farnham, *Saving Nature's Legacy: Origins of the Idea of Biological Diversity* (New Haven: Yale University Press, 2007), 1.

195 **"mediator, promoter, and catalyst"**: Farnham, *Saving Nature's Legacy*, 245.

196 **to call "ecosystem function"**: Previously called "ecosystem stability." Farnham, *Saving Nature's Legacy*, 181–92. For a report on the debate at its peak, see "Rift Over Biodiversity Divides Ecologists," *Science* 289 (2000): 1282–83.

196 **species play complementary roles within ecosystems**: See David Tilman, "Biodiversity & Environmental Sustainability Amid Human Domination of Global Ecosystems," *Daedalus* 141, no. 3 (2012): 108–20; and David Tilman et al., "Biodiversity and Ecosystem Functioning," *Annual Review of Ecology, Evolution, and Systematics* 45 (2014): 471–93.

196 **biodiversity is too broad and vague**: See, for example, Peter Kareiva and Michelle Marvier, "The Shortfalls of Biodiversity," *bioGraphic*, October 25, 2016.

197 **a rule of thumb for population viability**: Ian Robert Franklin, "Evolutionary Change

in Small Populations," and Michael Soulé, "Thresholds for Survival: Maintaining Fitness and Evolutionary Potential," in *Conservation Biology: An Evolutionary–Ecological Perspective*, edited by Michael Soulé and Bruce Wilcox (Sunderland, MA: Sinauer Associates, 1980), 135–70.

197 **Gilpin remembered Soulé's frustration**: Michael Gilpin, "Forty-Eight Parrots and the Origins of Population Viability Analysis," *Conservation Biology* 10, no. 6 (1996): 1491–93.

198 **a species of increasing concern**: Thomas R. Wellock, "The Dickey Bird Scientists Take Charge: Science, Policy, and the Spotted Owl," *Environmental History* 15, no. 3 (2010): 381–414.

198 **he and several students undertook a study**: Michael E. Soulé et al., "Reconstructed Dynamics of Rapid Extinctions of Chaparral-Requiring Birds in Urban Habitat Islands," *Conservation Biology* 2, no. 1 (1988): 75–92.

198 **as tropical ecologist William Laurance observed**: William F. Laurance, "Theory Meets Reality: How Habitat Fragmentation Research Has Transcended Island Biogeographic Theory," *Biological Conservation* 141 (2008): 1731–44.

199 **"national parks and equivalent reserves" more than doubled**: J. Harrison et al., "The World Coverage of Protected Areas: Development Goals and Environmental Needs," *Ambio* 11, no. 5 (1982): 239, figure 1.

199 **more than 10 percent of the land area**: Harrison et al., "The World Coverage of Protected Areas," 240, table 1.

199 **more than one hundred marine protected areas**: Sue Wells et al., "Building the Future of MPAs—Lessons From History," *Aquatic Conservation: Marine and Freshwater Ecosystems* 26, supplement 2 (2016): 106.

199 **displaced thousands of people**: Project Tiger, "Joining the Dots: The Report of the Tiger Task Force," Union Ministry of Environment and Forests, 2005, 89.

199 **1972 report called *The Limits to Growth***: Donella H. Meadows et al., *The Limits to Growth* (New York: Universe, 1972).

200 **the "Night of the Long Knives"**: Martin Holdgate, *The Green Web* (New York: Earthscan, 1999), 124. Also see Stephen J. Macekura, *Of Limits and Growth: The Rise of Global Sustainable Development in the Twentieth Century* (Cambridge: Cambridge University Press, 2015), 219–20.

201 **according to one prominent definition**: World Commission on Environment and Development, *Our Common Future* (Oxford: Oxford University Press, 1987), 8.

201 **The IUCN World Conservation Strategy**: John McCormick, "The Origins of the World Conservation Strategy," *Environmental Review* 10, no. 3 (1986): 177–87.

201 **"resisting all development"**: IUCN, "World Conservation Strategy: Living Resource Conservation for Sustainable Development," IUCN–UNEP–WWF, 1980.

201 **"the 'environment' is where we all live"**: Gro Bruntland, "Chairman's Foreword," in World Commission on Environment and Development, *Our Common Future*, ix.

202 **"the profane grail of Sustainable Development"**: Michael E. Soulé, "The Social Siege

of Nature," in *Reinventing Nature: Responses to Postmodern Deconstruction*, edited by
Michael E. Soulé and Gary Lease (Washington, DC: Island Press, 1995), 159.

202 **disparaged as "fortress conservation"**: John Hutton et al., "Back to the Barriers?
Changing Narratives in Biodiversity Conservation," *Forum for Development Studies*
32, no. 2 (2005): 341–70.

202 **"Parks cannot be maintained without order"**: John Terborgh, *Requiem for Nature*
(Washington, DC: Island Press, 2004), 192.

202 **"To change the fate of the world"**: Kent Redford and M. A. Sanjayan, "Retiring Cas-
sandra," *Conservation Biology* 17, no. 6 (2003): 1474.

203 **what they called "apex consumers"**: James A. Estes et al., "Trophic Downgrading of
Planet Earth," *Science* 333 (2011): 301–6.

204 **"You hold in your hands, I sincerely believe"**: Dave Foreman, "Around the Campfire," in *Wild Earth*, Special Issue: The Wildlands Project, 1992, 1.

204 **jobs "will be created, not lost"**: "The Wildlands Project Mission Statement," in *Wild
Earth*, Special Issue: The Wildlands Project, 1992, 3.

204 **"recognize conservation as one integral whole"**: Aldo Leopold, "Report to the American Game Conference on an American Game Policy," in *Transactions of the Seven-
teenth American Game Conference*, December 1–2, 1930, 287–88.

205 **financed by hunting license fees**: Frances Stead Sellers, "Hunting Is 'Slowly Dying
Off,' and That Has Created a Crisis for the Nation's Many Endangered Species,"
Washington Post, February 2, 2020.

205 **the backlash against conservation was lacerating**: Jonathan Thompson, "The First
Sagebrush Rebellion: What Sparked It and How It Ended," *High Country News*, January 14, 2016.

205 **practice a "politics of patience"**: Michael E. Soulé, "A Vision for the Meantime," in
Wild Earth, Special Issue: The Wildlands Project, 1992, 8.

205 **fired the imaginations of conservationists**: Charles C. Mann and Mark L. Plummer,
"The High Cost of Biodiversity," *Science* 260 (1993): 1868–71.

208 **"only by committing half of the planet's surface"**: E. O. Wilson, *Half-Earth: Our
Planet's Fight for Life* (New York: Liveright, 2016), 3.

209 **Critics of conservation biology**: See, for example, Peter Kareiva and Michelle Marvier, "What Is Conservation Science?" *BioScience* 62, no. 11 (2012): 962–69.

209 **reacted to these critiques with alarm**: See, for example, Michael E. Soulé, "The 'New
Conservation,'" *Conservation Biology* 27, no. 5 (2013): 895–97.

209 **An analysis of all of the research published**: Rogier E. Hintzen et al., "Relationship
Between Conservation Biology and Ecology Shown Through Machine Reading of
32,000 Articles," *Conservation Biology* 34, no. 3 (2020): 721–32.

210 **suggests philosopher Christine Korsgaard**: Christine M. Korsgaard, *Fellow Creatures: Our Obligations to the Other Animals* (Oxford: Oxford University Press, 2018),
48–50.

211 **"the worst thing we could do in Ethiopia"**: Quoted in Steve Chase, ed., *Defending the*

Earth: A Dialogue Between Murray Bookchin and Dave Foreman (Boston: South End Press, 1991), 108. Foreman originally said this in 1986, during an interview published in the Australian magazine *Simple Living*. The interviewer was Bill Devall, co-author of the book *Deep Ecology*. In *Defending the Earth*, Foreman prefaced the statement by saying, "I have been insensitive, albeit unintentionally, and for that I humbly apologize" (107). He added: "I still have honest questions about the much-admired relief effort during the Ethiopian famine . . . it has to be asked, and I admit it is a terrible question, if such last-minute relief efforts actually allow a human population stretched beyond the land's carrying capacity to eke out existence for a few more years and, in the process, cause even greater deterioration of the land's capacity to support humans and other species" (110–11).

211   **"the illusion that we can escape the troubles of the world"**: William Cronon, "The Trouble with Wilderness, or Getting Back to the Wrong Nature," in *Uncommon Ground: Rethinking the Human Place in Nature*, edited by William Cronon (New York: W. W. Norton, 1996), 69–90. This widely cited and still controversial essay, worth reading in its entirety, argues that while the "myth of wilderness" suggests that "we can somehow leave nature untouched by our passage," our challenge is instead "to decide what kind of marks we wish to leave" (88).

212   **"conservation is a human endeavor"**: Michael B. Mascia et al., "Conservation and the Social Sciences," *Conservation Biology* 17, no. 3 (2003): 650.

212   **"ask for the help of professionals"**: Soulé, "Conservation Biology and the 'Real World,'" 9.

212   **"based on the premise that industrialization"**: Aldo Leopold, "Memo, On Professor Ecological Economics," Aldo Leopold Papers, University of Wisconsin Archives, series 10–1, box 3, folder 8, 533–34.

213   **His favored candidate**: Qi Feng Lin, "Aldo Leopold's Unrealized Proposals to Rethink Economics," *Ecological Economics* 108 (2014): 104–14.

213   **"One of the anomalies of modern ecology"**: Aldo Leopold, "Wilderness ('The two great cultural advances . . .')," in *Aldo Leopold: A Sand County Almanac and Other Writings on Ecology and Conservation*, edited by Curt Meine (New York: Library of America, 2013), 375.

213   **attempting to quantify the myriad benefits**: For an overview of current research on ecosystem services and natural capital, see Stephen Polasky et al., "Role of Economics in Analyzing the Environment and Sustainable Development," *Proceedings of the National Academy of Sciences* 116, no. 2 (2019): 5233–38, and associated colloquium papers.

213   **For Leopold, quantifying the public benefits**: Lin, "Aldo Leopold's Unrealized Proposals to Rethink Economics." Today, there is an interdisciplinary field called ecological economics, and its transformative goals overlap with Leopold's. However, it has struggled to distinguish itself from the more conventional discipline of envi-

ronmental economics. See Robert L. Nadeau, "The Unfinished Journey of Ecological Economics," *Ecological Economics* 109 (2015): 101–8.

CHAPTER EIGHT: THE RHINO AND THE COMMONS

217 **"No one should shoot one of my elephants!"**: Garth Owen-Smith, *An Arid Eden: A Personal Account of Conservation in the Kaokoveld* (Johannesburg: Jonathan Ball, 2010), 387.

220 **an escalating power struggle**: Marion Wallace, *A History of Namibia: From the Beginning to 1990* (New York: Columbia University Press, 2011), 273–308.

220 **a new generation of conservationists was emerging**: William Adams and David Hulme, "Conservation and Community: Changing Narratives, Policies and Practices in African Conservation," in *African Wildlife and Livelihoods: The Promise and Performance of Community Conservation*, edited by David Hulme and Marshall Murphree (Oxford: James Currey, 2001), 9–23.

221 **support from subsistence hunters and farmers was often critical**: William Adams et al., "Biodiversity Conservation and the Eradication of Poverty," *Science* 306 (2004): 1146–49.

221 **simple cost-sharing arrangements**: Edmund Barrow et al., "The Evolution of Community Conservation Policy and Practice," in *African Wildlife and Livelihoods*, 59–72.

221 **a short-lived but influential accord**: David Western, "Ecosystem Conservation and Rural Development: The Case of Amboseli," in *Natural Connections: Perspectives in Community-Based Conservation*, edited by David Western and Michael Wright (Washington, DC: Island Press, 1994).

222 **Huxley . . . had supported utilization**: Julian Huxley, "Cropping the Wild Protein," *Observer*, November 20, 1960, 23.

222 **the tragedy of the commons**: Garrett Hardin, "The Tragedy of the Commons," *Science* 162 (1968): 1243–48.

222 **Elinor Ostrom, a political scientist at Indiana University**: Vlad Tarko, *Elinor Ostrom: An Intellectual Biography* (London: Rowman and Littlefield, 2017), 76.

222 **The common features of these systems**: Thomas Dietz et al., "The Struggle to Govern the Commons," *Science* 302 (2003): 1907–12.

223 **"We are neither trapped in inexorable tragedies"**: Elinor Ostrom, "A Behavioral Approach to the Rational Choice Theory of Collective Action: President's Address, American Political Science Association, 1997," *American Political Science Review* 92, no. 1 (1998): 16.

224 **"When someone told you"**: Elinor Ostrom, "The Challenge of Reforming Resource Governance," address to the South Asian Network for Development and Environmental Economics, December 7, 2011.

224 " 'What is *the* way of doing something?' ": Ostrom, "The Challenge of Reforming Resource Governance."

224 **what she called "panaceas"**: Ostrom, "The Challenge of Reforming Resource Governance."

224 **In southern Africa in the late 1980s**: Adams and Hulme, "Conservation and Community," 9–23.

225 **"It is easy for us who have full stomachs"**: Owen-Smith, *An Arid Eden*, 415.

227 **conducted her research from a dung-plastered hut**: See Margaret Jacobsohn, *Life Is Like a Kudu Horn: A Conservation Memoir* (Johannesburg: Jacana Media, 2019).

227 **"they also want power"**: Owen-Smith, *An Arid Eden*, 495.

228 **what was essentially a war zone**: Wallace, *A History of Namibia*, 273–308.

230 **more than eighty communal conservancies in Namibia**: Statistics on Namibian conservancies are collected annually by the Namibian Association of CBNRM Support Associations and available at nacso.org.na.

233 **as varied as the people who undertake them**: William Adams and David Hulme, "If Community Conservation Is the Answer in Africa, What Is the Question?" *Oryx* 35, no. 3 (2001): 193–200.

234 **"What we have ignored"**: Kenneth J. Arrow et al., "An Uncommon Woman for the Commons," *Proceedings of the National Academy of Sciences* 109, no. 33 (2012): 13135.

234 **hundreds of community-based conservation efforts**: See, for example, Jeremy S. Brooks et al., "How National Context, Project Design, and Local Community Characteristics Influence Success in Community-Based Conservation Projects," *Proceedings of the National Academy of Sciences* 109, no. 52 (2012): 21265–70.

236 **the continent's vulture species**: Darcy Ogada et al., "Another Continental Vulture Crisis: Africa's Vultures Collapsing Toward Extinction," *Conservation Letters* 9, no. 2 (2016): 89–97.

236 **bulked up their arsenals in response**: For more on the militarization of park rangers in South Africa and throughout the African continent, see Cathleen O'Grady, "The Price of Protecting Rhinos," *Atlantic*, January 13, 2020.

239 **at the annual TED conference**: John Kasaona, "How Poachers Became Caretakers," talk at TED2010, February 13, 2010, ted.com.

239 **Roosevelt standing next to a fallen elephant**: This elephant was one of hundreds of animals killed by Roosevelt and his son Kermit on their 1909 safari, as enumerated by Theodore Roosevelt in *African Game Trails: An Account of the African Wanderings of an American Hunter–Naturalist* (London: John Murray, 1910), 437–39. The trip was co-sponsored by the Smithsonian, and the massive haul was ostensibly for science; many of the carcasses became museum specimens. For a list of the casualties, see Phil Edwards, "All 512 animals Teddy Roosevelt and his son killed on safari," *Vox*, February 3, 2016.

239 **can bring long-simmering conflicts . . . to a roaring boil**: For a more measured but still heated academic discussion of the issue, see Chelsea Batavia et al., "The Ele-

phant (Head) in the Room," and Amy J. Dickman et al., "Is There an Elephant in the Room? A Response to Batavia et al.," both in *Conservation Letters* 12, no. 1 (2018).

240 **at a Smithsonian Institution conservation conference**: "Working with Communities," panel at the Smithsonian Earth Optimism Summit, April 22, 2017, video at earthoptimism.si.edu.

240 **rows of sandy mounds**: See David Olusoga and Casper W. Erichsen, *The Kaiser's Holocaust: Germany's Forgotten Genocide* (London: Faber and Faber, 2010), 255–56. Also see Wallace, *A History of Namibia*, 155–82.

240 **people who died in the prison camps**: These atrocities were first exposed to the outside world in the 1918 "Blue Book," a document assembled by British officials but later suppressed throughout the British Empire. See Jeremy Silvester and Jan-Bart Gewald, eds., *"Words Cannot Be Found": German Colonial Rule in Namibia: An Annotated Reprint of the 1918 Blue Book* (Leiden: Brill, 2003), xxx–xxxii. Not until late 2019 did German government officials acknowledge that the colonial "war" had been a genocide.

240 **forerunners of the Nazi concentration camps**: Olusoga and Erichsen, authors of *The Kaiser's Holocaust*, are among the proponents of the "continuity thesis," which argues that the German labor camps in Namibia were direct ancestors of the Nazi concentration camps.

242 **those who have lost loved ones and livelihoods**: For examples, see Max Bearak, " 'I Hate Elephants': Behind the Backlash Against Botswana's Giants," *Washington Post*, June 7, 2019.

CHAPTER NINE: THE FEW WHO SAVE THE MANY

248 **the planet has lost**: Based on the species in the "Extinct" category of the IUCN Red List of Threatened Species, iucnredlist.org.

248 **one extinction per million species per year**: Stuart L. Pimm et al., "The Future of Biodiversity," *Science* 269 (1995): 347–50.

248 **one thousand extinctions per million species per year**: Jurriaan M. De Vos, "Estimating the Normal Background Rate of Species Extinction," *Conservation Biology* 29, no. 2 (2015): 452–62.

249 **close to or already entering the sixth extinction**: Gerardo Ceballos et al., "Accelerated Modern Human-Induced Species Losses: Entering the Sixth Mass Extinction," *Science Advances* 1, no. 5, e1400253.

249 **as a "biological annihilation"**: Gerardo Ceballos et al., "Biological Annihilation via the Ongoing Sixth Mass Extinction Signaled by Vertebrate Population Losses and Declines," *Proceedings of the National Academy of Sciences* 114, no. 30 (2017): E6089–96. Also see Ceballos et al., "Vertebrates on the Brink as Indicators of Biological Annihilation and the Sixth Mass Extinction," *Proceedings of the National Academy of Sciences* 117, no. 24 (2020): 13596–602.

249 **A 2016 study of more than eight thousand . . . species:** Sean Maxwell et al., "Biodiversity: The Ravages of Guns, Nets, and Bulldozers," Comment, *Nature* 546 (2016): 143–45.

250 **In 2019, a global assessment:** Sandra Diaz et al., "Pervasive Human-Driven Decline of Life on Earth Points to the Need for Transformative Change," *Science* 366 (2019): eaax3100.

250 **slowed vertebrate species' tumble toward extinction:** Michael Hoffmann et al., "The Impact of Conservation on the Status of the World's Vertebrates," *Science* 330 (2010): 1503–9.

250 **a "Green List" of recovered species:** Leslie Evans Ogden, "A New Green List: Metrics for Recovery," *BioScience* 69, no. 2 (2019): 156.

252 **lost nearly ten million acres of forest:** Melissa Leach and James Fairhead, "Challenging Neo-Malthusian Deforestation Analyses in West Africa's Dynamic Landscapes," *Population and Development Review* 26, no. 1 (2000): 29. Leach and Fairhead suggest a revision of previous estimates, which they argue persuasively were too high. In recent years, deforestation in Ghana and the rest of West Africa has accelerated; for more, see Emmanuel Opoku Acheampong, "Deforestation Is Driven by Agricultural Expansion in Ghana's Forest Reserves," *Scientific African* 5 (2019): e00146.

254 **less attention to mutually beneficial relationships, called mutualisms:** Judith L. Bronstein, "The Study of Mutualism," in *Mutualism*, edited by Judith L. Bronstein (Oxford: Oxford University Press, 2015), 3–19.

254 **measured in scarcity, or calories, or tourism potential:** Many have attempted to enumerate, and quantify, the tangible and intangible benefits of different species to human society. For one example, see Claude Gascon et al., "The Importance and Benefits of Species," *Current Biology* 25 (2015): R431–38.

254 **to a large extent "inscrutable to us":** Rachel Carson, *The Edge of the Sea* (Boston: Houghton Mifflin, 1998), 250. Originally published in 1955.

255 **healthy fish populations relieve hunting pressure:** Justin S. Brashares et al., "Bushmeat Hunting, Wildlife Declines, and Fish Supply in West Africa," *Science* 306 (2004): 1180–83.

255 **"Nature advocates have obtained much":** Holly Doremus, "The Rhetoric and Reality of Nature Protection: Toward a New Discourse," *Washington and Lee Law Review* 57, no. 1 (2000): 14.

256 **"maintain the population of wild flora and fauna":** Article 2 of the Convention on the Conservation of European Wildlife and Natural Habitats (a.k.a. the Bern Convention), text republished by the International Environmental Agreements Database Project, iea.uoregon.edu.

256 **Parties to the 1992 Convention on Biological Diversity:** Text of the convention republished by the International Environmental Agreements Database Project, iea

.uoregon.edu. The "ecosystem approach" was adopted at the Conference of the Parties in 1995.

256 **few concrete benefits for biodiversity:** Piero Visconti et al., "Protected Area Targets Post-2020," *Nature* 364 (2019): 239–41.

256 **a loss of some three billion birds:** Kenneth V. Rosenberg et al., "Decline of the North American Avifauna," *Science* 366 (2019): 120–24.

257 **"This little thought experiment":** Christopher D. Stone, *Should Trees Have Standing?: Law, Morality, and the Environment* (Oxford: Oxford University Press, 2010), xi.

258 **the Supreme Court ruled against the Sierra Club:** *Sierra Club v. Morton,* 405 U.S. 727 (1972). The defendant in the case was the Secretary of the Interior, so when the office changed hands the defendant changed, too.

259 **"If Justice Douglas has his way":** Quoted in Stone, *Should Trees Have Standing,* xiv.

259 **"We thought that we would":** Quoted in Stone, *Should Trees Have Standing,* xv.

259 **"more tender and widely diffused":** Charles Darwin, *The Descent of Man and Selection in Relation to Sex,* vol. 1 (New York: D. Appleton, 1872), 99.

259 **in 2020, the U.S. Fish and Wildlife Service proposed to severely curtail:** Lisa Friedman, "Birds Will Be Next Casualties of Regulatory Rollback," *New York Times,* January 31, 2020, B4. In August 2020, a U.S. District Court judge struck down the rule changes.

260 **noted in a 2010 article:** James D. K. Morris and Jacinta Ruru, "Giving Voice to Rivers: Legal Personality as a Vehicle for Recognising Indigenous Peoples' Relationships to Water?," *Australian Indigenous Law Review* 14, no. 2 (2010): 58.

260 **In 2014, the parties reached a settlement:** For the settlement and other documents, see ngatangatatiaki.co.nz.

260 **"like our most fundamental societal stories":** Doremus, "Rhetoric and Reality of Nature Protection," 45.

260 **the Whanganui legislation and a growing number of laws like it:** See, for example, David R. Boyd, *The Rights of Nature: A Legal Revolution that Could Save the World* (Toronto: ECW Press, 2017), 223–27.

261 **writes that the Māori worldview:** Mere Roberts et al., "Kaitiakitanga: Maori Perspectives on Conservation," *Pacific Conservation Biology* 2, no. 1 (1995): 16.

261 **"a universal symbiosis":** Aldo Leopold, "The Conservation Ethic," in *Aldo Leopold: A Sand County Almanac and Other Writings on Ecology and Conservation,* edited by Curt Meine (New York: Library of America, 2013), 333.

CONCLUSION: *HOMO AMPHIBIUS*

262 **a subspecies related to the southwestern black rhinos:** Southwestern black rhinos are one of the four (three extant) subspecies of black rhino; southern white rhinos are one of the two subspecies of white rhinos. White rhinos and black rhinos are

related. The fourth species of black rhino, the western black rhino, was declared extinct in 2011.

262 **the first white rhino born via artificial insemination in North America**: The storied history of white rhinos at the San Diego Zoo begins in 1971, when twenty rhinos spent almost two weeks en route from South Africa to California by tramp steamer and train. Edward is the ninety-ninth white rhino born at the Safari Park. For more, see Douglas G. Meyers, *Mister Zoo: The Life and Legacy of Dr. Charles Schroeder* (San Diego: Zoological Society of San Diego, 1999), 235–39; and Ian Player, *The White Rhino Saga* (New York: Stein and Day, 1973).

263 **hope to "rewind" the extinction**: Joseph Saragusty et al., "Rewinding the Process of Mammalian Extinction," *Zoo Biology* 35, no. 4 (2016): 280–92.

263 **looks like an oversized dental saliva ejector**: Yes, there is a video. Matt Simon, "The Plan to Save the Rhino with a Cervix-Navigating Robot," *Wired*, April 18, 2019.

263 **the "Pets" section of *People***: Helen Murphy, "Southern White Rhino Calf Conceived through Artificial Insemination Born at San Diego Zoo," *People*, August 1, 2019.

263 **"hopeful interventions"**: For more on Ryder's work and its underlying philosophy, see his essay "Opportunities and Challenges for Conserving Small Populations: An Emerging Role for Zoos in Genetic Rescue," in *The Ark and Beyond: The Evolution of Zoo and Aquarium Conservation*, edited by Ben A. Minteer et al. (Chicago: University of Chicago Press, 2018), 255–66.

263 **the captive breeding and release of the California condor**: Like the "high-intensity period of 'Poultry Husbandry'" proposed for the whooping crane, the plan to capture and breed the last surviving California condors, proposed in the mid-1970s, was strenuously opposed by many conservationists. Paul Ehrlich, in a letter to the Secretary of the Interior in 1980, called it "nothing more than unconscionable and dangerous harassment." See John Nielsen, *Condor: To the Brink and Back: The Life and Times of One Giant Bird* (New York: Harper Perennial, 2006), 145–48.

264 **they all know that years, perhaps decades, of costly work remain**: Oliver Ryder et al., "Exploring the Limits of Saving a Subspecies: The Ethics and Social Dynamics of Restoring Northern White Rhinos (*Ceratotherium simum cottoni*)," *Conservation Science and Practice* (2020), e241.

265 **often described human beings as amphibians**: A preoccupation explored by R. S. Deese in *We Are Amphibians: Julian and Aldous Huxley on the Future of Our Species* (Berkeley: University of California Press, 2014).

265 **"Whether we like it or not, we are amphibians"**: Aldous Huxley, "The Education of an Amphibian," in *Tomorrow and Tomorrow and Tomorrow and Other Essays* (New York: Signet, 1964), 13.

266 **"We're standing on the cusp of a new era"**: Jennifer A. Doudna and Samuel H. Sternberg, *A Crack in Creation: Gene Editing and the Unthinkable Power to Control Evolution* (New York: Houghton Mifflin Harcourt, 2017), 243.

266 **"We are as gods and might as well get used to it"**: Stewart Brand, "Purpose" statement of *The Whole Earth Catalog*, 1968. In his book *Whole Earth Discipline: An Ecopragmatist Manifesto* (New York: Viking, 2009), Brand adjusted his maxim to "we are as gods and have to get good at it."

266 **"We of the genus *Homo* ride the logs"**: Aldo Leopold, "The Round River," in *A Sand County Almanac with Essays on Conservation from Round River* (New York: Ballantine Books, 1991), 188 and 196. In an earlier typescript of this essay, the second part of the quote is a rhetorical question: "Are we burling our log of state with skill, or only with energy?" Apparently, either Leopold or a posthumous editor settled on the answer before publication.

267 **the 1946 edition of *Brave New World***: Aldous Huxley, *Brave New World* (New York: Harper and Brothers, 1946).

267 **offered his character a third alternative**: Huxley elaborated on his vision of this alternative in his final novel *Island* (New York: Harper Perennial, 2009). Originally published in 1962.

268 **"Alone among the animals"**: Margaret Atwood, "'Everybody Is Happy Now,'" *Guardian*, November 17, 2007.

268 **Sociologist Carrie Friese speculates**: From a talk given at "Expanding the Discussion: Ethical and Social Issues in the Northern White Rhino Genetic Rescue Initiative," a conference held on October 14 and 15, 2019, at the San Diego Zoo Institute for Conservation Research.

268 **"affected by the same environmental influences"**: Rachel Carson, "The Pollution of Our Environment," in *Lost Woods: The Discovered Writing of Rachel Carson*, edited by Linda Lear (Boston: Beacon Press, 1998), 30–32.

269 **"man with all his noble qualities"**: Charles Darwin, *The Descent of Man and Selection in Relation to Sex*, vol. 2 (London: John Murray, 1871), 405.

Further Reading

In following the connections among the people, ideas, and institutions that make up the conservation ecosystem, I could touch only briefly on some tantalizing chapters in conservation history. Happily, many are explored in depth elsewhere.

CHAPTER ONE: THE BOTANIST WHO NAMED THE ANIMALS

Wilfrid Blunt, *Linnaeus: The Compleat Naturalist* (London: Frances Lincoln, 2004), first published in 1971, is still the definitive biography of Linnaeus in English. Also helpful is *Linnaeus: The Man and His Work*, edited by Tore Frängsmyr (Berkeley: University of California Press, 1983). Andrea Wulf, *The Brother Gardeners: Botany, Empire, and the Birth of An Obsession* (New York: Vintage, 2010) follows Linnaeus and other eighteenth-century naturalists on their often frenzied quests to expand the known universe of plants.

Linnaeus's *Philosophia Botanica*, which includes his bossy instructions to other taxonomists, has been beautifully translated into English by Stephen Freer (Oxford: Oxford University Press, 2003).

For more on the enduring urge to classify, see Paul Lawrence Farber, *Finding Order in Nature: The Naturalist Tradition from Linnaeus to E. O. Wilson* (Baltimore: Johns Hopkins University Press, 2000); Harriet Ritvo, *The Platypus and the Mermaid, and Other Figments of the Classifying Imagination* (Cambridge, MA: Harvard University Press, 1997); and Carol Yoon, *Naming Nature: The Clash Between Instinct and Science* (New York: W. W. Norton, 2009).

The story of Archaea and its discovery is told in David Quammen, *The Tangled Tree: A Radical New View of Life* (New York: Simon and Schuster, 2018). Ernst Mayr's *The*

Growth of Biological Thought: Diversity, Evolution, and Inheritance (Cambridge, MA: Belknap Press of Harvard University Press, 1982) is a classic reference for anyone interested in the study of life and the humans who undertake it.

CHAPTER TWO: THE TAXIDERMIST AND THE BISON

William Hornaday is the subject of at least two popular biographies: Gregory Dehler, *The Most Defiant Devil: William Temple Hornaday and His Controversial Crusade to Save American Wildlife* (Charlottesville, VA: University of Virginia Press, 2013), and Stefan Bechtel, *Mr. Hornaday's War: How a Peculiar Victorian Zookeeper Waged a Lonely Crusade for Wildlife That Changed the World* (Boston: Beacon Press, 2012).

For more on bison, people, and their shared history, see Andrew C. Isenberg, *The Destruction of the Bison: An Environmental History, 1750–1920* (Cambridge: Cambridge University Press, 2000), and the anthology *Bison and People on the North American Great Plains: A Deep Environmental History*, edited by Geoff Cunfer and Bill Waiser (College Station: Texas A&M University Press, 2016).

The history of species extinction as an idea is documented by Mark V. Barrow Jr. in his excellent book *Nature's Ghosts: Confronting Extinction from the Age of Jefferson to the Age of Ecology* (Chicago: University of Chicago Press, 2009). For more on George Perkins Marsh, who contained multitudes, see David Lowenthal, *George Perkins Marsh: Prophet of Conservation* (Seattle: University of Washington Press, 2003).

The historical intersections of conservation and racism are explored by Miles A. Powell in *Vanishing America: Species Extinction, Racial Peril, and the Origins of Conservation* (Cambridge, MA: Harvard University Press, 2016), and by Jonathan Peter Spiro in his engrossing if unfortunately titled *Defending the Master Race: Conservation, Eugenics, and the Legacy of Madison Grant* (Lebanon, NH: University of Vermont Press, 2009).

Ota Benga's own story, overlooked for too long, is told by Pamela Newkirk in *Spectacle: The Astonishing Life of Ota Benga* (New York: Amistad, 2015).

For more on Theodore Roosevelt's conservation legacy, see Douglas Brinkley, *The Wilderness Warrior: Theodore Roosevelt and the Crusade for America* (New York: Harper Perennial, 2010); *Theodore Roosevelt: Naturalist in the Arena*, edited by Char Miller and Clay S. Jenkinson (Lincoln, NE: University of Nebraska Press, 2020); and Keith Aune and Glenn Plumb, *Theodore Roosevelt and Bison Restoration on the Great Plains* (Charleston, SC: History Press, 2019).

Douglas Coffman's search for the Hornaday bison is recounted in his memoir *Reflecting the Sublime: The Rebirth of an American Icon* (Fort Benton, MT: River & Plains Society, 2013).

CHAPTER THREE: THE HELLCAT AND THE HAWKS

Dyana Furmansky's entertaining and thorough biography *Rosalie Edge, Hawk of Mercy: The Activist Who Saved Nature from the Conservationists* (Athens, GA: University of Georgia Press, 2009), draws on family correspondence not available elsewhere.

John James Audubon is the subject of several biographies, including Richard Rhodes, *John James Audubon: The Making of an American* (New York: Alfred A. Knopf, 2004), and William Souder, *Under a Wild Sky: John James Audubon and the Making of the Birds of America* (Minneapolis: Milkweed Editions, 2014).

For more on George Bird Grinnell and his lifelong dedication to conservation, see John Taliaferro, *Grinnell: America's Environmental Pioneer and His Restless Drive to Save the West* (New York: Liveright, 2019).

The central role of bird enthusiasts in the early history of conservation is detailed in Mark V. Barrow Jr., *A Passion for Birds: American Ornithology After Audubon* (Princeton: Princeton University Press, 1998), and Frank Graham Jr., *The Audubon Ark: A History of the Audubon Society* (Austin: University of Texas Press, 1990).

Maurice Broun recalled his years on North Lookout in his memoir *Hawks Aloft!: The Story of Hawk Mountain* (New York: Dodd, Mead, 1948).

The history of the animal-welfare movement, including the key role of women in its development, is explored in Janet M. Davis, *The Gospel of Kindness: Animal Welfare and the Making of Modern America* (Oxford: Oxford University Press, 2016).

CHAPTER FOUR: THE FORESTER AND THE GREEN FIRE

The most comprehensive Leopold biography is the wonderfully readable *Aldo Leopold: His Life and Work* by Curt Meine (Madison: University of Wisconsin Press, 2010). Also very valuable are Susan L. Flader, *Thinking Like a Mountain: Aldo Leopold and the Evolution of an Ecological Attitude Toward Deer, Wolves, and Forests* (Madison: University of Wisconsin Press, 1994), and Julianne Lutz Newton, *Aldo Leopold's Odyssey: Rediscovering the Author of A Sand County Almanac* (Washington, DC: Island Press, 2006). The Library of America's *Aldo Leopold: A Sand County Almanac and Other Writings on Ecology and Conservation*, edited by Curt Meine and published in 2013, includes previously unpublished selections from Leopold's letters and journals.

Estella B. Leopold's *Stories from the Leopold Shack: Sand County Revisited* (Oxford: Oxford University Press, 2016) is a charming, poignant memoir of family and place.

For many years, the massacre of Lakota Sioux by U.S. soldiers at Wounded Knee Creek has been portrayed as the final spiritual and cultural defeat of Native America. For a thorough repudiation of this idea, see David Treuer, *The Heartbeat of Wounded Knee* (New York: Riverhead, 2019).

John Muir's epic life is explored in the suitably epic *A Passion for Nature: The Life of John Muir* by Donald Worster (Oxford: Oxford University Press, 2008). Char Miller's *Gif-*

ford Pinchot and the Making of Modern Environmentalism (Washington, DC: Island Press, 2001) rescues the man from the caricature.

For more on conservation in Germany before and during the Third Reich, see Frank Uekotter, *The Green and the Brown: A History of Conservation in Nazi Germany* (Cambridge: Cambridge University Press, 2006). Julia Boyd, *Travelers in the Third Reich: The Rise of Fascism, 1919–1945* (New York: Pegasus Books, 2018) unearths the eyewitness accounts and varied reactions of those who, like Leopold, visited Germany during Hitler's ascent.

William Beebe, though mentioned only in passing here, was famous in his time for his nervy scientific adventures, many of which are related by Carol Grant Gould in *The Remarkable Life of William Beebe* (Washington, DC: Island Press, 2004).

Donald Worster's *Nature's Economy: A History of Ecological Ideas* (Cambridge: Cambridge University Press, 1994) is an indispensable history of ecology, beginning with its first conceptual glimmerings and continuing through its development as a scientific discipline.

CHAPTER FIVE: THE PROFESSOR AND THE ELIXIR OF LIFE

The collective literary output of the Huxley family is formidable, and ranges from the dazzling to the best forgotten. While Julian Huxley's blueprints for society have not aged well, his two-part memoir (*Memories I and II*, London: Allen and Unwin, 1970 and 1973) is revealing and often admirably self-aware. Juliette Huxley's memoir *Leaves of the Tulip Tree* (London: John Murray, 1986) complicates and deepens Julian's self-portrait.

A fine introduction to the shy, brilliant, self-doubting man who was Charles Darwin is David Quammen, *The Reluctant Mr. Darwin: An Intimate Portrait of Charles Darwin and the Making of His Theory of Evolution* (New York: W. W. Norton, 2006).

Two useful overviews of the internationalization of the conservation movement and its close relationship with colonialism are William Adams, *Against Extinction: The Story of Conservation* (London: Earthscan, 2004) and Rachelle Adam, *Elephant Treaties: The Colonial History of the Biodiversity Crisis* (Lebanon, NH: University Press of New England, 2014). I also learned a great deal from *Conservation in Africa: People, Politics, and Practice*, edited by David Anderson and Richard Grove (Cambridge: Cambridge University Press, 1987), an academic volume published during a time of transition in international conservation.

The bewildering landscape of international environmental agreements is made comprehensible by *Lyster's International Wildlife Law*, intended as a textbook but accessible to general readers. The revised and updated second edition (Cambridge: Cambridge University Press, 2011) was edited by Michael Bowman et al.

For more on the eugenics movement and its long aftermath, which extends into the present day, see Alexandra Minna Stern, *Eugenic Nation: Faults and Frontiers of Better Breeding in Modern America* (Oakland: University of California Press, 2016).

The intriguing life, important work, and execrable opinions of William Vogt are detailed in Charles C. Mann, *The Wizard and the Prophet: Two Remarkable Scientists and Their Dueling Visions to Shape Tomorrow's World* (New York: Vintage, 2018), which contrasts Vogt with Green Revolution architect Norman Borlaug.

For more on Vogt and Leopold's intellectual mutualism, see the unexpectedly absorbing *Guano and the Opening of the Pacific World: A Global Ecological History* by Gregory T. Cushman (Cambridge: Cambridge University Press, 2013).

CHAPTER SIX: THE EAGLE AND THE WHOOPING CRANE

For many years *Rachel Carson: Witness for Nature* by Linda Lear (Boston: Mariner Books, 1997) was the only full-fledged Carson biography, but it is now bookended by William Souder, *On a Farther Shore: The Life and Legacy of Rachel Carson* (New York: Penguin Random House, 2012).

For much more on the history of DDT, see David Kinkela, *DDT and The American Century: Global Health, Environmental Politics, and the Pesticide that Changed the World* (Chapel Hill: University of North Carolina Press, 2011). Also useful is *DDT, Silent Spring, and the Rise of Environmentalism: Classic Texts*, edited by Thomas Dunlap (Seattle: University of Washington Press, 2008).

Robert Porter Allen, who was a delightful writer as well as a dedicated field ornithologist, recounted his adventures in his memoirs *The Flame-Birds* (New York: Dodd, Mead, 1947) and *On the Trail of Vanishing Birds* (New York: McGraw–Hill, 1957). Kathleen Kaska tells his life story in *The Man Who Saved the Whooping Crane* (Gainesville: University Press of Florida, 2012).

The Birds of Heaven: Travels with Cranes by Peter Matthiessen (New York: Farrar, Straus and Giroux, 2003) follows George Archibald on his perpetual journey to protect cranes worldwide. Archibald's own anecdotes, collected in *My Life with Cranes: A Collection of Stories* (Baraboo, WI: International Crane Foundation, 2016) capture both his sunny idealism and remarkable determination.

Charles F. Wurster, *DDT Wars: Rescuing Our National Bird, Preventing Cancer, and Creating the Environmental Defense Fund* (Oxford: Oxford University Press, 2015), is an important reminder that *Silent Spring* was only the beginning of the hard work that led to the ban on DDT.

Paul Warde, Libby Robin, and Sverker Sörlin, *The Environment: A History of the Idea* (Baltimore: Johns Hopkins University Press, 2018), is a brief but illuminating exploration of a concept often assumed to have no history.

On the history of the Endangered Species Act, two succinct and solid sources are Shannon Petersen, *Acting for Endangered Species: The Statutory Ark* (Lawrence: University

Press of Kansas, 2002), and Steven Lewis Yaffee, *Prohibitive Policy: Implementing the Federal Endangered Species Act* (Cambridge, MA: MIT Press, 1982).

The University of Tennessee law professor who sued over the snail darter is Zygmunt J. B. Plater, and his book about the experience is *The Snail Darter and the Dam: How Pork Barrel Politics Endangered a Little Fish and Killed a River* (New Haven: Yale University Press, 2014).

CHAPTER SEVEN: THE SCIENTISTS WHO ESCAPED THE TOWER

What are sometimes called the "founding documents" of conservation biology are collected in *Conservation Biology: An Evolutionary–Ecological Perspective*, edited by Michael Soulé and Bruce Wilcox (Sunderland, MA: Sinauer Associates, 1980), and *Conservation Biology: The Science of Scarcity and Diversity*, edited by Michael Soulé (Sunderland, MA: Sinauer Associates, 1986). Also see *Collected Papers of Michael E. Soulé: Early Years in Modern Conservation Biology* (Washington, DC: Island Press, 2014).

The estimable Victor Shelford is lifted from obscurity by Robert Croker in *Pioneer Ecologist: The Life and Work of Victor Ernest Shelford, 1877–1968* (Washington, DC: Smithsonian Institution Press, 1991).

Paul Sabin's book *The Bet: Paul Ehrlich, Julian Simon, and Our Gamble Over Earth's Future* (New Haven: Yale University Press, 2013) uses the Ehrlich–Simon wager as an opportunity to delve into the thinking of both men. Thomas Robertson, *The Malthusian Moment: Global Population Growth and the Birth of American Environmentalism* (New Brunswick, NJ: Rutgers University Press, 2012) follows the history of human population growth as an environmental issue in the United States, from William Vogt and Fairfield Osborn to Paul Ehrlich and into the 1980s.

John Kricher's *The Balance of Nature: Ecology's Enduring Myth* (Princeton: Princeton University Press, 2009) is a useful corrective to one of the most stubborn popular misunderstandings about ecology. Timothy J. Farnham, *Saving Nature's Legacy: Origins of the Idea of Biological Diversity* (New Haven: Yale University Press, 2007) and David Takacs, *The Idea of Biodiversity: Philosophies of Paradise* (Baltimore: Johns Hopkins University Press, 1996), examine the history and ambiguities of conservation's dominant concept.

Mary Ellen Hannibal tells the story of the Wildlands Project and its ambitious vision in *The Spine of the Continent: The Race to Save America's Last, Best Wilderness* (Guilford, CT: Lyons Press, 2013).

Stephen J. Macekura's *Of Limits and Growth: The Rise of Global Sustainable Development in the Twentieth Century* (Cambridge: Cambridge University Press, 2015) follows the winding path of the international conservation movement from its beginnings in colonial Africa through its embrace of—and eventual disillusionment with—the goal of sustainable development.

CHAPTER EIGHT: THE RHINO AND THE COMMONS

Garth Owen-Smith's memoir *An Arid Eden: A Personal Account of Conservation in the Kaokoveld* (Johannesburg: Jonathan Ball, 2010) details his decades of innovative work in his favorite place on earth. Margaret Jacobsohn, *Life Is Like a Kudu Horn: A Conservation Memoir* (Johannesburg: Jacana Media, 2019) recalls both the high adventures and everyday satisfactions of a life in conservation.

For more about the lives and culture of the Himba people, see Margaret Jacobsohn, *Himba: Nomads of Namibia*, which includes wonderful photos by Peter Pickford (Johannesburg: Penguin Random House South Africa, 1998).

Raymond Bonner's *At the Hand of Man: Peril and Hope for Africa's Wildlife* (New York: Knopf, 1993), though almost thirty years old, is still relevant as a perceptive and thoroughly reported look at the contradictions of the international conservation establishment's work in Africa.

Elinor Ostrom's career is distilled in Vlad Tarko, *Elinor Ostrom: An Intellectual Biography* (London: Rowman and Littlefield, 2017). For an overview of her common property research, see her book *Governing the Commons: The Evolution of Institutions for Collective Action* (Cambridge: Cambridge University Press, 1990).

David Western, one of the first practitioners of community-based conservation, recalls his trials, errors, and successes in *In the Dust of Kilimanjaro* (Washington, DC: Island Press, 1997). Other early projects are documented in *Natural Connections: Perspectives in Community-Based Conservation*, edited by David Western and R. Michael Wright (Washington, DC: Island Press, 1994).

For more on Namibia, a young nation with a long history, see Marion Wallace, *A History of Namibia: From the Beginning to 1990* (New York: Columbia University Press, 2011), and David Olusoga and Casper W. Erichsen, *The Kaiser's Holocaust: Germany's Forgotten Genocide* (London: Faber and Faber, 2010).

Andrew Loveridge, one of the zoologists who spent nearly a decade studying the Zimbabwean lion named Cecil, writes about Cecil's killing and its significance for wildlife conservation in *Lion Hearted: The Life and Death of Cecil and the Future of Africa's Iconic Cats* (New York: Regan Arts, 2018).

CHAPTER NINE: THE FEW WHO SAVE THE MANY

Elizabeth Kolbert's *The Sixth Extinction: An Unnatural History* (New York: Henry Holt, 2014) is essential reading for anyone interested in the past and future of species conservation.

Christopher D. Stone's *Should Trees Have Standing? Law, Morality, and the Environment* (Oxford: Oxford University Press, 2010) includes his original *Southern California Law Review* article and his reflections on its impact. David R. Boyd, *The Rights of*

Nature: A Legal Revolution That Could Save the World (Toronto: ECW Press, 2017) is an enthusiastic overview of ongoing innovations in environmental law.

For a beautiful meditation on the concept of reciprocity between humans and the rest of life, see Robyn Wall Kimmerer, *Braiding Sweetgrass: Indigenous Wisdom, Scientific Knowledge, and the Teachings of Plants* (Minneapolis: Milkweed Editions, 2013).

CONCLUSION: *HOMO AMPHIBIUS*

R. S. Deese, *We Are Amphibians: Julian and Aldous Huxley on the Future of Our Species* (Berkeley: University of California Press, 2014), uses the Huxley brothers' favorite metaphor to examine their respective intellectual paths and their influences on each other.

Jennifer Doudna's *A Crack in Creation: Gene Editing and the Unthinkable Power to Control Evolution* (Boston: Mariner Books, 2017) is a valuable view of our possible futures.

Island, Aldous Huxley's final novel (New York: Harper Perennial, 2009), is a very odd book, but fascinating in light of his regrets about the dichotomy he offered in *Brave New World*. I recommend it.

Illustration Credits

129   Lady Ottoline Morrell / © National Portrait Gallery, London
146   © Terence Spencer / The LIFE Images Collection via Getty Images / Getty
      Images
149   Shirley A. Briggs / © Rachel Carson Council
174   Courtesy of the International Crane Foundation
186   Gene Arias / © 1979 NBCUniversal / Getty Images
208   © JT Thomas
218   © John Dambik / Alamy Stock Photo
223   Courtesy Lilly Library, Indiana University, Bloomington, Indiana
238   © Jeff Muntifering
243   Karin le Roux / Courtesy Margaret Jacobsohn
246   © Todd Pusser
253   © Emma Hutlin
264   Ken Bohn / © San Diego Zoo Safari Park

Index

Page numbers in *italics* refer to illustrations.